THE ENCYCLOPEDIA OF
ANIMAL ECOLOGY

THE ENCYCLOPEDIA OF ANIMAL ECOLOGY

Edited by Dr Peter D. Moore

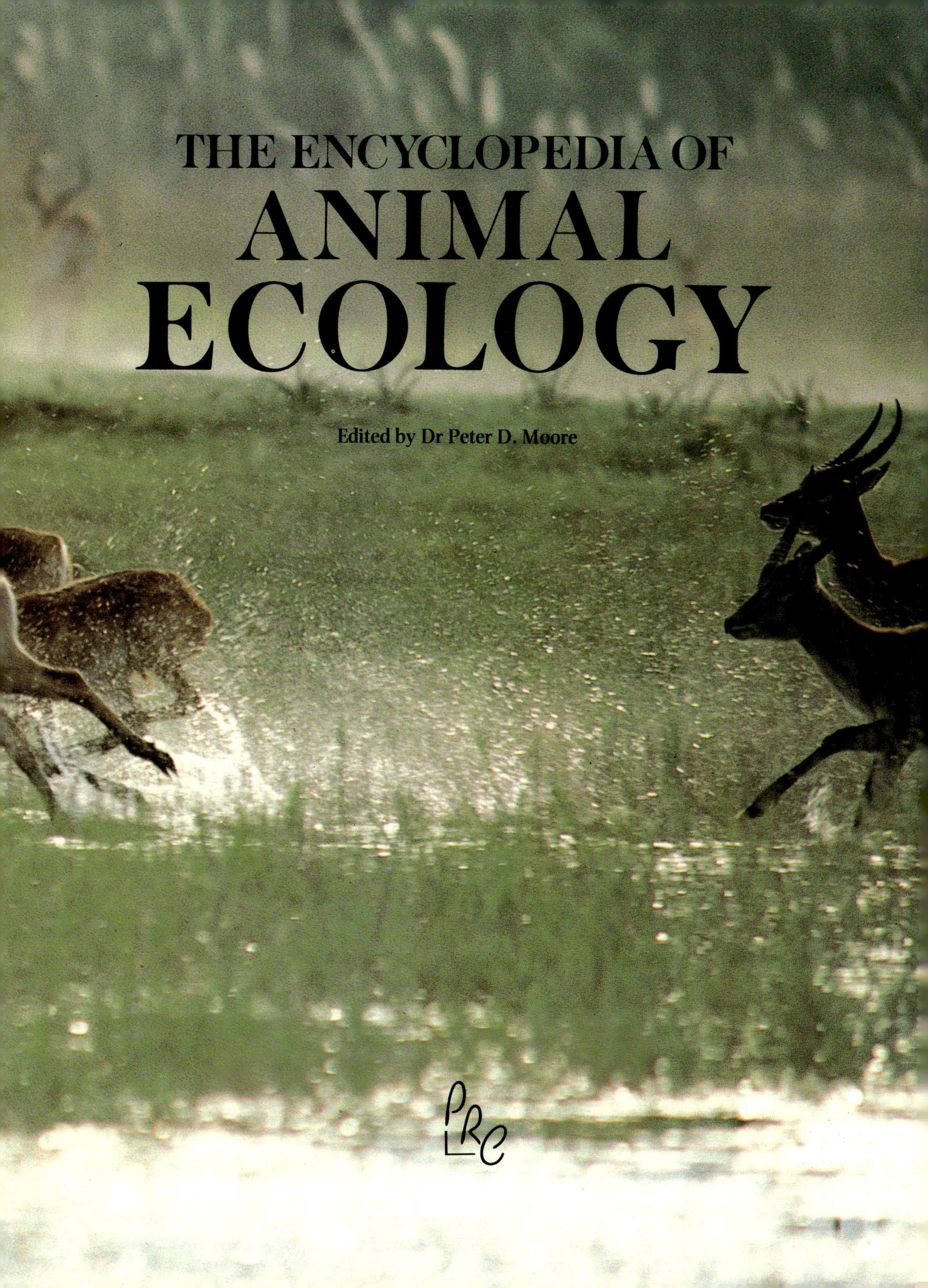

PRC

AN EQUINOX BOOK

Planned and produced by:
Equinox (Oxford) Ltd
Musterlin House,
Jordan Hill, Oxford,
England OX2 8DP

This edition published 1991 by The Promotional Reprint Company Limited

ISBN 1 85648 012 7

Printed and bound in Hong Kong

Advisory Editor

Dr Paul A. Colinvaux
The Ohio State University,
Columbus, Ohio,
USA

Artwork Panels

Denys Ovenden

Roger Gorringe

Richard Lewington

Mick Loates

Contributors

GEB Gary E. Belovsky
The University of Michigan
Ann Arbor, Michigan
USA

RAB Richard A. Brown
ICI Plant Protection Division
Bracknell
England

CBC C. Barry Cox
King's College London
England

RHE Roland H. Emson
King's College London
England

NF Nigel Franks
University of Bath
England

BG Barrie Goldsmith
University College London
England

OG Olavi Gronwall
University of Stockholm
Sweden

DWM David W. Macdonald
University of Oxford
England

PDM Peter D. Moore
King's College London
England

JO Jennifer Owen
Leicester
England

SS Steven Smith
King's College London
England

BDT Brian D. Turner
King's College London
England

BW Bernd Wursig
Moss Landing Marine
Laboratories California
USA

*Left: American marten (*Martes americana*) in tree (Leonard Lee Rue III); half-title: Blue-spotted stingray (*Taeniura lymma*) (Oxford Scientific Films); title page: lechwe (*Kobus leche*) galloping across flooded grassland in Okavango delta, Botswana (NHPA).*

PREFACE

Ecology is the art of seeing the whole as well as the pieces; it is the ability to see the wood as well as the trees. Ecology is a very young science, having emerged as a distinct discipline from the work of the great natural historians of the last two centuries, including that of Charles Darwin. By the beginning of this century, biology was extending its frontiers in many directions, at the microscopic level into the structure of the cell, at a chemical level in the understanding of cell biochemistry, at the genetic level and in the area of the physiology of whole organisms. The time was ripe for the development of a new discipline which could fuse the discoveries of biology with the observations of natural history. The result was the emergence of ecology.

Ecology takes the whole animal or plant, the organism, which is itself a collection of cells often gathered into distinct organs, and studies it in relation to its environment. The environment of each individual is very complex, for it includes the physical and chemical components of the surroundings, such as climate, acidity, salinity, etc, as well as the influence of all of the other organisms in the vicinity. These may include other individuals of the same species, which may be competing for the limited resources of the site, as well as other individuals of other species. These other species can represent a hazard or an opportunity. They may be predators, parasites or pathogens, in which case their presence is clearly harmful. Or they may be food species, or host species, or may simply provide a suitable habitat, as does a reed bed for a bittern, or the hair of a sloth for the algae which coat it.

Because of the complexity of interactions in the natural world, ecologists find it necessary to conduct their studies at a variety of levels. Some take a single species and study a group of individuals from it, which they term a population. They may study the tolerance of these individuals to extremes of various factors, or their optimum requirements, the temperature, humidity or light intensity at which they do best; or they may examine the interactions between individuals when they are in competition for limited resources.

Ecologists also need to know about breeding rates of populations and life expectancy so that they can understand how effectively the population can maintain itself under various conditions. Clearly such studies are of great importance in the management and conservation of the wildlife of the earth, whether we are considering the herring stocks of the Atlantic Ocean or the survival of the Giant panda.

Some ecologists concern themselves with the ways in which two or more species populations interact with one another. Do they feed upon one another, in which case how does the predator affect the abundance of the prey? Or, conversely, if the prey becomes scarce does this affect the population density of the predator? Some species have very similar life-styles and may find themselves in competition, in which case the ecologist needs to know which species is more efficient under a range of different conditions. The species which comes out best in a competitive struggle in one set of conditions, be they cold, wet, disturbed, and so on, may not succeed if the conditions are hot or dry or more stable.

In the real world we often find large numbers of species, all interacting together against a background of an almost infinite range of physical and chemical variables and combinations. Here, ecologists are dealing with the community of living organisms in their physical setting and it is possible to study the ways in which its composition (in terms of species) and its structure (that is, the architecture of its vegetation) varies in the course of time or when the community is subjected to a new stress, such as pollution.

A final level of study for the ecologist is the ecosystem. Here, the community of living organisms and their non-living setting are taken together and studies are made of the ways in which chemical elements migrate in and out and around the whole system. They may move, like carbon, from atmosphere to plant, to herbivore, to carnivore, to fungus, to soil and back to the atmosphere in a never ending cycle. In the ecosystem it is also possible to study the flow of energy from the sunlight, through plants, animals and microbes eventually to be dissipated as heat. Here is one of the most satisfying models of ecology, in which we see all of the specific components functioning together as an organic whole.

This volume sets out to describe and explain the many ideas and concepts which ecologists have developed to describe and explain the world of nature, and much of the first section of the book is devoted to this purpose. Many of the principles briefly mentioned in this introduction are expanded and discussed in detail and are illustrated by specific examples from nature.

But this is not a textbook of theoretical ecology. Rather it seeks to take the principles of ecology and show how they apply to real, practical situations. In the second section of the book the major climatic zones of the world and their resident fauna and flora are described, and it is shown how the ecological principles outlined in the first section apply in practice in determining which species can occur in each area and how the communities thus formed relate to their physical environment and function as an ecosystem.

In this second section, attention is not confined to the communities of the land, the so-called terrestrial biomes, but the very important and influential aquatic habitats are also covered, including both fresh and salt water. These are not true biomes in the sense that their biological composition is determined primarily by climatic zones, but they form convenient units in which conditions are very much more uniform than in their terrestrial counterparts.

The third and final section of the book describes some of the specific problems facing human beings in their ecological setting. Our species has constructed communities with the quite deliberate aim of chanelling energy into its own development—a process we call agriculture. We have also constructed new, artificial habitats for ourselves—the cities. These gross modifications of natural systems have provided an opportunity for a range of organisms to take advantage of the new conditions and exploit them accordingly. Since these creatures often find themselves in competition with us, we term them "pests" and we seek methods, both chemical and biological, for their control. In doing so we create new problems not only for the pests, but for ourselves.

The recognition that the human species is a component of the global ecosystem rather than a totally independent unit, has greatly influenced social and political thinking in recent

Green python (*Chondropython viridis*) in Australian rain forest (Frithfoto).

decades. In particular, it has caused us to develop a new and rational attitude to the many species with which we share our planet. The sentiments of the poet John Donne apply here: we are not an island, but part of a greater whole, and any loss of part of that whole is a loss to ourselves. Donne's emphasis was on the unity of humankind, but the thought is equally true of the entire body of our planet. Damage or loss to any component part is bound to have repercussions which will be felt by the whole—including ourselves.

Each section of this volume is introduced by a summary, followed by a series of main entries written by an eminent team of authors; sometimes entries have been written by two authors, enabling us to benefit from their differing expertise. On many occasions topics worthy of special treatment are presented as boxed features or even double-page special features. Ecology is about animals and their habitats, so we have been very careful to select photographs of animals depicted in their natural surroundings. To these have been added many diagrams and composite color artwork panels showing examples of animals typical of a particular habitat or region.

Much careful design and thought has gone into the construction of this book and credit for this is due to the staff at Equinox (Oxford) Ltd, particularly Dr Graham Bateman. Theirs was the initial idea behind the work and theirs has been the hard task of checking detail and obtaining the best possible art and photographic materials. The message of this book is an important one and it is appropriate that it should be presented in an attractive and a convincing form. If this has been achieved, then it is due largely to the efforts of the Equinox team.

Peter D. Moore
DEPARTMENT OF BIOLOGY
KING'S COLLEGE (KQC)
UNIVERSITY OF LONDON

CONTENTS

Butterflies, Belenois aurota, *drinking on damp ground on migration; South Africa (Premaphotos Wildlife).*

Themes of Ecology

OUR appreciation of the living world is greatly enhanced when we begin to understand how different animals and plants relate to one another. Why, for example, do certain animals and plants often occur together in distinct communities? Do they all simply prefer the same type of environmental conditions, or are they in some respect dependant upon one another for their continued survival? Both explanations may be true for various species and it is the task of ecologists to sort out the complex web of inter-relationships which we find in nature.

In this first section we shall consider all the different pieces of information which we may need in order to explain why a particular animal or plant is found where it is and what role it is playing on nature's stage. We need to know about the responses of each species both to the physical setting and to the other participants in the interactions of life. And how does the community function as a whole? These constitute some of the great themes of ecology.

◄ **Curled in a burrow,** European rabbits (*Oryctolagus cuniculus*) grooming in their underground shelter. In some circumstances, for example of high population levels, rabbits live all their lives above ground; myxomatosis, spread by fleas in burrows, may have encouraged this behavior.

WHAT IS ECOLOGY?

What is ecology?... No two individuals are alike... Populations in their environment... Competition to survive... Interaction between species... Nature as an ecosystem...

THE natural world is a complex whole. To study animal and plant species in isolation is therefore somewhat artificial. Each has its own place in the overall pattern of nature, and the task of ecology is to investigate how the various species interact with each other and with their environment.

Ecology, then, is the study of living things in relation to their *surroundings*. By "surroundings" is meant not only the non-living aspects of their habitat, such as the soil, air, climate etc comprising the physical environment, with its chemical and physical properties. The surroundings also include the other living organisms which interact with one another in a range of complex ways. One cannot really understand why an organism has a particular structure, or a certain biochemical peculiarity or behavior pattern, unless it is studied in the surroundings within which it evolved. Only then can one appreciate how it fits into the pattern of nature. Such study forms the basis of the science of ecology.

The first level of study in ecology is the *individual*. Anyone who has worked closely with animals realizes that no two individuals are precisely alike. For example, they vary in their response to stimuli, and react in different ways in any given situation. Much of this variation is genetically based (ie inherited) and forms the raw material upon which natural selection operates in the evolutionary process. Those physical characteristics and those behavioral responses which render an individual more likely to survive and to breed will tend to persist among its progeny, while those characters which render an individual less fit for survival tend to be eliminated.

This individual variation, although so important in the process of evolution, creates certain problems for the biologist. One should never study a single individual and assume that the studies will be representative of the species. Instead one must study a *population*, which is a collection of individuals of a given species. The term can be applied at any scale, from tadpoles in a pool to herrings in the Atlantic Ocean. Within the population, one would expect a range of structure, physiology and behavioral response which provides a more complete picture of the nature of the animal (or plant) under study. One can ascertain the optimal requirements of a population by studying how various individuals respond to a range of conditions. One can thus discover the preferred conditions of temperature, humidity, salinity, acidity, light and so on, which may well assist in understanding why a particular species is found in a certain habitat. A woodlouse, for example, is sensitive to water loss, and hence it is normally found under conditions of high humidity, in forest leaf litter or under stones. Lizards are less susceptible to desiccation but need to bask in the sun to warm up and enhance their metabolic rates, so they are often found in open, sunny sites. Some plants, such as rhododendrons, can only grow on acid soils because of the chemical nature of the membranes bounding their cells. So a great deal can be understood about the physical and chemical needs of a species by a careful study of populations.

A further factor can be introduced by considering the interactions between members of a population. Since these all have very similar requirements, such as food, space, preferred sites and so on, they will be in competition with one another should any of these resources be in short supply. Such competition may result in the death of some individuals. Tadpoles in a small pond and birch saplings in a forest clearing are both engaged in a process of mutual interaction where the resources are adequate to support only a few adults; the remainder must die. Here again, the process of natural selection acts like a sieve and ensures that only the very fittest survive.

But the real world is more complex than this simple picture of a population of individuals interacting with one another against a physical environmental setting. There are other populations of other species, all adapted to make a living in their various ways from the same area. Each species will influence the way of life of others around it. The tree in a forest modifies the local climate as it casts shade, encloses humid air masses among its foliage and provides shelter from high temperature and wind. In turn, its growth may be influenced by the insects which feed upon its foliage or bore into its bark, the yeasts which gain energy from the sugars leaking out of

▼ **Levels of study in ecology**—diagram illustrating the structure of a woodland ecosystem. When the community of living animals and plants is put into its physical, chemical, meteorological and geological setting, the whole can be viewed as a complex dynamic mechanism with its specialized parts. The ecologist may study the whole complex system, or a population of one species, or a single individual.

► **Lone puma.** The individual is the first level of study in ecology. No two individuals are ever alike. Variation in individuals is the raw material upon which natural selection and hence evolution works. Shown here is a puma (*Felis concolor*), a species of many habitats, from forest to desert, throughout the Americas.

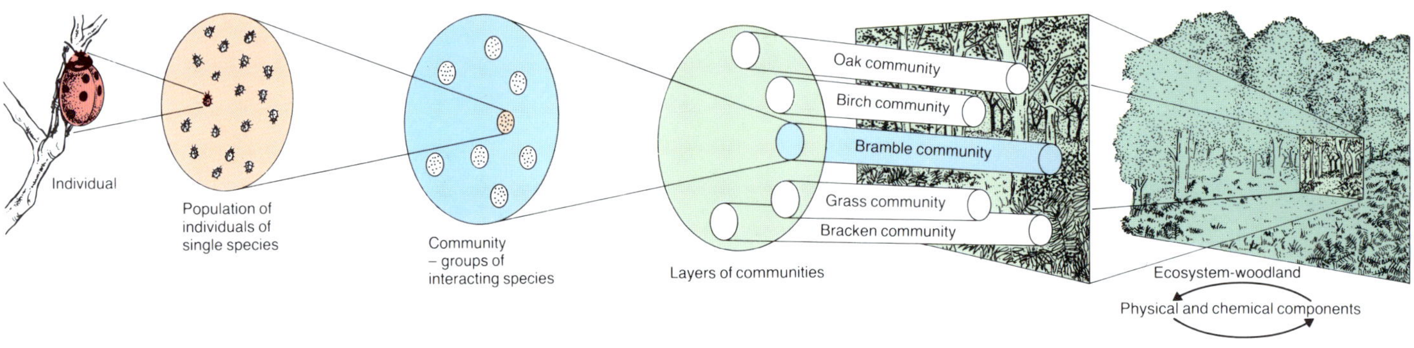

its leaves, the pathogenic fungi which invade its wounds and parasitize the living tissues, the decomposers in the soil which release nutrients from the litter around its roots, and the squirrels and pigs which feed upon its fruits but which may also play a part in their dispersal. No species is an island, but each is influenced by all other species around it. This interacting assemblage of populations of various species can be looked upon as a *community*. Some communities, such as those inhabiting glacial rubble or shingle beaches, are clearly unstable. In these change takes place until eventually an equilibrium between the various component populations is achieved. Others are stable in the sense that over the limited span of a human life little overall change is observed, as in the case of most forests; species interlock in their resource requirements, and therefore the rate of extinction of species from the community is reduced to virtually zero. Over geological time, however, no community can be regarded as stable.

When the community of living plants and animals is put into its physical, chemical, meteorological and geological setting, the whole can be viewed as a complex, dynamic mechanism with its specialist component parts. The green plants are fixing energy which is consumed by grazing animals, which are in turn fed upon by predators and parasites. Dead tissues fall to the floor, where an army of detritivores (consumers of debris) and decomposers, both animal and microbe, rely upon their residual energy as the means to a livelihood. The chemical minerals of the rocks are weathered to release elements needful for the maintenance of life, and these are efficiently extracted by plant roots to become available either to herbivore or detritivore, according to their fate, only to be released again in an endless cycle.

This orderly structure is the *ecosystem*, and is arguably the most realistic and practically useful way in which the world of nature can be conceived. PDM

ENERGY AND NUTRIENT RELATIONS

The need for energy... How plants absorb and store energy... Plant-eating animals... Meat-eating animals... Transfer of energy through feeding... Trophic (feeding) levels... The need for nutrients... Different organisms have different requirements... Animals need sodium, plants do not... Where are nutrients to be found?... How plants and animals make use of them... The global cycle, and how man has changed it... The nutrient cycle in individual ecosystems...

LIVING organisms function in many ways like machines and, like machines, they need fuel in order to operate. Every movement of an animal consumes energy, so one of the most fundamental questions in ecology is how this energy can be supplied. The one unfailing available source of energy is the sun, and one of the early steps in the evolution of life must have been the development of a biochemical technique whereby solar energy could be captured and stored.

The primitive organisms which developed this technique evolved into the group now known as plants. Those which preyed upon or parasitized these energy fixers formed the line of evolution leading to animals and to a large proportion of the microbes, the bacteria and the fungi.

Animals that feed on living plant tissues are collectively called *herbivores*. Most plant material contains large amounts of cell-wall material, mainly cellulose. Because this is difficult to consume, herbivores tend to have basically similar mouthparts irrespective of what animal group they come from. Such mouthparts consist of two elements, a cutting or incisor region and a broad ridged area for grinding. The teeth of cows, sheep and horses, for example, can be readily identified with this pattern, and the mandibles of locusts and caterpillars have a sharp edge and a grinding area at their base. Almost all invertebrate herbivores (such as locusts) pass cellulose through their digestive systems virtually untouched. Some insects, such as aphids, avoid cellulose by feeding directly on the plant's sap. But among the mollusks there are species which have in their digestive system an enzyme called cellulase which splits cellulose into simple sugars. Species which appear to feed on almost pure cellulose, termites and wood-boring beetles, cannot themselves digest it but rely on single-celled animals (protozoa), bacteria

▼ **Predators and prey**—a pair of dingoes (*Canis dingo*) combine to attack a Lace monitor (*Varanus varius*) at a water hole in New South Wales. Monitors are themselves voracious predators, but here the tables are turned.

► **Power from the sun**—a diagram summarizing the basic processes of photosynthesis (see box right).

Nature's Powerhouse—Photosynthesis

The process of photosynthesis whereby plants manage to tap the sun's energy is complex, but it consists essentially of two stages. First, a pigment (green chlorophyll) absorbs some of the energy of sunlight and this activates it in such a way that part of the energy is captured chemically in the formation of a substance known as ATP. This acts as a transporter and temporary store of energy in the cell. The reaction also results in the production of a chemical reducing agent, called NADPH; both of these compounds are needed at a later stage. But in this process the chlorophyll molecules lose negatively-charged particles (electrons), and a convenient and abundant source of these subatomic particles is water. The extraction of electrons from water causes the release of oxygen as an unwanted but very important byproduct.

But long-term storage of energy requires more stable compounds than the ATP carrier produced by this process, so a second stage begins which does not itself require light. In this, carbon dioxide is taken into the plant and is reduced to organic compounds in a reaction that uses energy, supplied by the ATP, and reducing power supplied by NADPH, both generated in the chlorophyll/light reaction. The organic compounds produced, which are various types of sugars, may be used for building new tissues, for food storage, or for respiration to generate the energy needed to keep the plant cells operating. The new plant tissues produced in photosynthesis are rich in energy and form a resource which animals are able to exploit, directly (eg herbivores) or indirectly (carnivores).

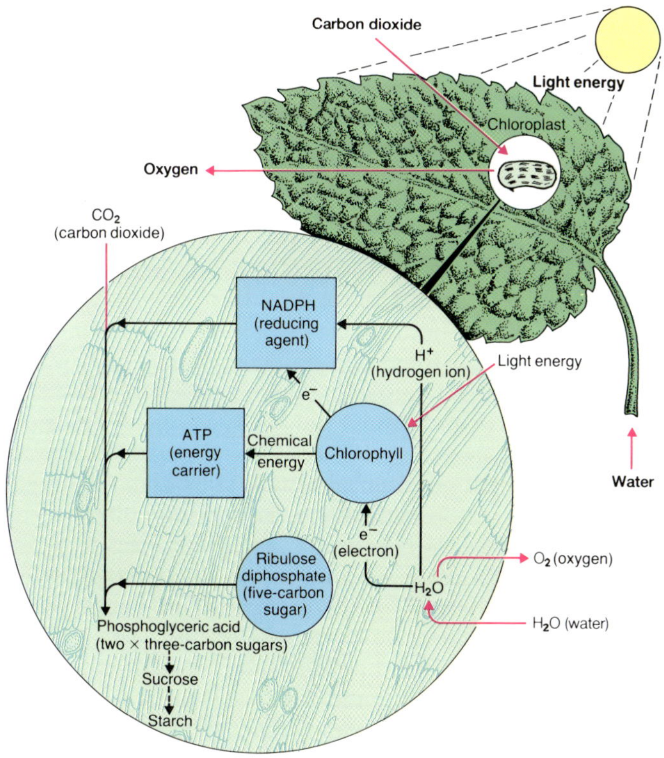

and fungi in their guts to break it down to sugars which can be absorbed. Most herbivorous mammals also use microorganisms, housed in specially developed areas of the gut, to process cellulose.

There are several types of meat-eating animal. Predators catch, kill and consume living animals, while carrion feeders eat the leftovers of predators, and also animals newly dead from natural causes or through disease or accident. Large predatory mammals, such as lions, are rarely successful in catching healthy prey. They usually eat the old, the sick or the very young, and so they tend to have little effect on the population size and age structure of their prey, only feeding on those which would die anyway. Invertebrate predators—spiders, wasps, dragonflies and so on—are far more effective at getting their prey. They often use highly toxic stings and venoms or powerful jaws to deal with even strong and healthy prey. These predators have a marked controlling influence on the population structure of their prey.

Feeding is, of course, the way in which animals get the energy they need to exist. Energy passes through a community via feeding, or *trophic*, links. This interconnection of species on the basis of what feeds on what is called a *foodweb*. Species are also grouped together on the basis of what they feed on. These groupings are called *trophic levels*. Plants form the first, producer, trophic level; then come the consumer levels, the herbivores, carnivores and parasites. So within the overall food-web, distinct trophic levels can be identified, containing species with similar feeding habits.

From studying foodwebs it has been found that the number of trophic levels is limited to a maximum of five or six. Several explanations of this interesting fact have been proposed. First, animals tend to eat things smaller than themselves, so higher trophic levels would contain larger and larger animals, and this may lead to size limitations. Secondly, of the energy entering a trophic level only a small proportion is available to be eaten by the next trophic level. This is because much of the energy taken in by animals is used by them in their daily life. The number of trophic levels may thus be limited by the availability of energy. If this is so, there should be a greater number of trophic levels in those areas where there is more primary productivity, for example in rain forests. This does not seem to be so. There is no relationship between primary productivity and the number of trophic levels. The most recent suggestion, based on computer simulations, is that complex communities with many trophic levels would be unstable and incapable of responding to small disturbances. Without the ability to tolerate change, communities collapse.

Animal and plant nutrition consists of more than the simple acquisition of energy from photosynthesis or through the foodwebs. When ecologists talk of nutrients, they refer to those chemical elements which are involved in the construction of living tissues and which are therefore needed by both plants and animals. The most important of these in terms of bulk requirements are carbon, hydrogen and oxygen, but many other elements are also required, such as nitrogen, potassium, calcium, sulfur, phosphorus, etc. Plants obtain their nutrients either from the atmosphere (carbon and oxygen) or from the soil (nitrogen, potassium, calcium etc, which are extracted via the roots). Animals are dependent upon their food and water intake for their nutrient supply as well as for the satisfaction of their energy needs.

There are, however, situations in which an animal needs to supplement its normal diet with additional elements, especially where soils, and therefore vegetation, are poor in these. One must bear in mind also that an animal's need for elements is not identical with that of a plant. Take calcium, for example. Plants need calcium so that the membranes that bound their cells will function efficiently, but mammals need much larger supplies of this element because of its importance in skeletal structure. So animals may find themselves short of calcium in areas where it is scarce, such as heathland and moorland. It is not only mammals that are affected in this way; those gastropods (such as snails) which construct shells have a high calcium requirement and are not generally found on heathlands. Gastropods lacking shells (slugs) are not restricted in this way and are a common element of the heath fauna.

An even more curious problem is found in the case of sodium supply. Plant physiologists are generally agreed that this element is not essential to plants, although most contain a little. Most plants, in fact, expend considerable amounts of energy in absorbing potassium, which they need, and excluding sodium, which they do not need, but which closely resembles potassium in its chemical properties. Animals, on the other hand, do require sodium, mainly because of its involvement in nerve function. This disparity in requirements can lead to sodium deficiency among herbivores, especially if they are unable to supplement their diet by licking sodium-rich rocks or soil (see box, p10).

When the chemical elements needed by living creatures are considered on a global scale, it is found that most are reasonably abundant, although they may occupy a variety of reservoirs. Oxygen, for example, occurs as a gas in the atmosphere, dissolved in the oceans (hydrosphere) and chemically combined in the rocks (lithosphere). Carbon is present as carbon dioxide in rather low concentrations in the atmosphere (0.03 percent by volume), as bicarbonate ions in the hydrosphere and combined as carbonates in the lithosphere. Nitrogen is an unusual element in that it is an abundant component of the atmosphere (79 percent) and yet the majority of organisms cannot obtain it directly from this source. Since it is an important component of living tissues, especially as proteins, it is vital that nitrogen should be available to plants and animals. Fortunately, there are some microbial organisms which have the biochemical equipment to change the somewhat inert nitrogen gas to ammonium compounds and hence to incorporate it into the living components of the planet (the biosphere). These organisms include free-living creatures, such as the bacterial genus *Nitrosomonas* and the blue-green algae (strictly bacteria, the Cyanobacteria), and also some organisms which are symbiotically associated with another species, such

Interactions in a Lake Community

A foodweb is a description of the feeding connections between the organisms that make up a community. Energy passes through the community, from one trophic level to the next, via these foodweb links. To a greater or lesser degree all communities have a similar trophic structure. By way of example consider a lake-dwelling community (see RIGHT).

Solar energy is utilized by the green plants which grow in the shallow peripheral water and by the microscopic phytoplankton floating in the surface layers of the lake. These plants form the first trophic level of primary producers. Plants are eaten by herbivores.

Small planktonic animals such as daphnia and cyclops, collectively referred to as *zooplankton*, filter the phytoplankton from the water. The water plants and the coating of microscopic plant (and animal) life are grazed by mollusks, mayfly larvae, and other plant-eating insects. Some fish, such as the Grass carp and the bitterling, also feed predominantly on plants. This diverse collection of animals comprises the second trophic level, the herbivores.

The third trophic level is equally varied and comprises the carnivores that feed on herbivores. The larvae of many insects, dragonflies, water beetles, and some caddis flies are carnivorous and eat the smaller herbivores. Zooplankton form the staple diet of several fish.

The fourth trophic level sometimes blurs into the third. Animals occupying the fourth level consume carnivores, although during their juvenile stages they may feed on herbivores. Again a diverse group of species occupies this level. Many of the insectivorous fish feed on insects which may be herbivores or carnivores. Trout, bream, carp, roach and rudd all fall into this category.

The fifth trophic level contains birds such as the heron and kingfisher, and fishermen that catch and eat trout, carp and bream. Pike and perch are also fourth- or fifth-level feeders; and many, by eating these fish (especially perch) may be feeding at a sixth level.

In this simple example it is easy to see that species cannot always be placed definitely in particular trophic levels. The age and size of the species, and the availability of food items, are important factors in determining where an animal is positioned in the foodweb.

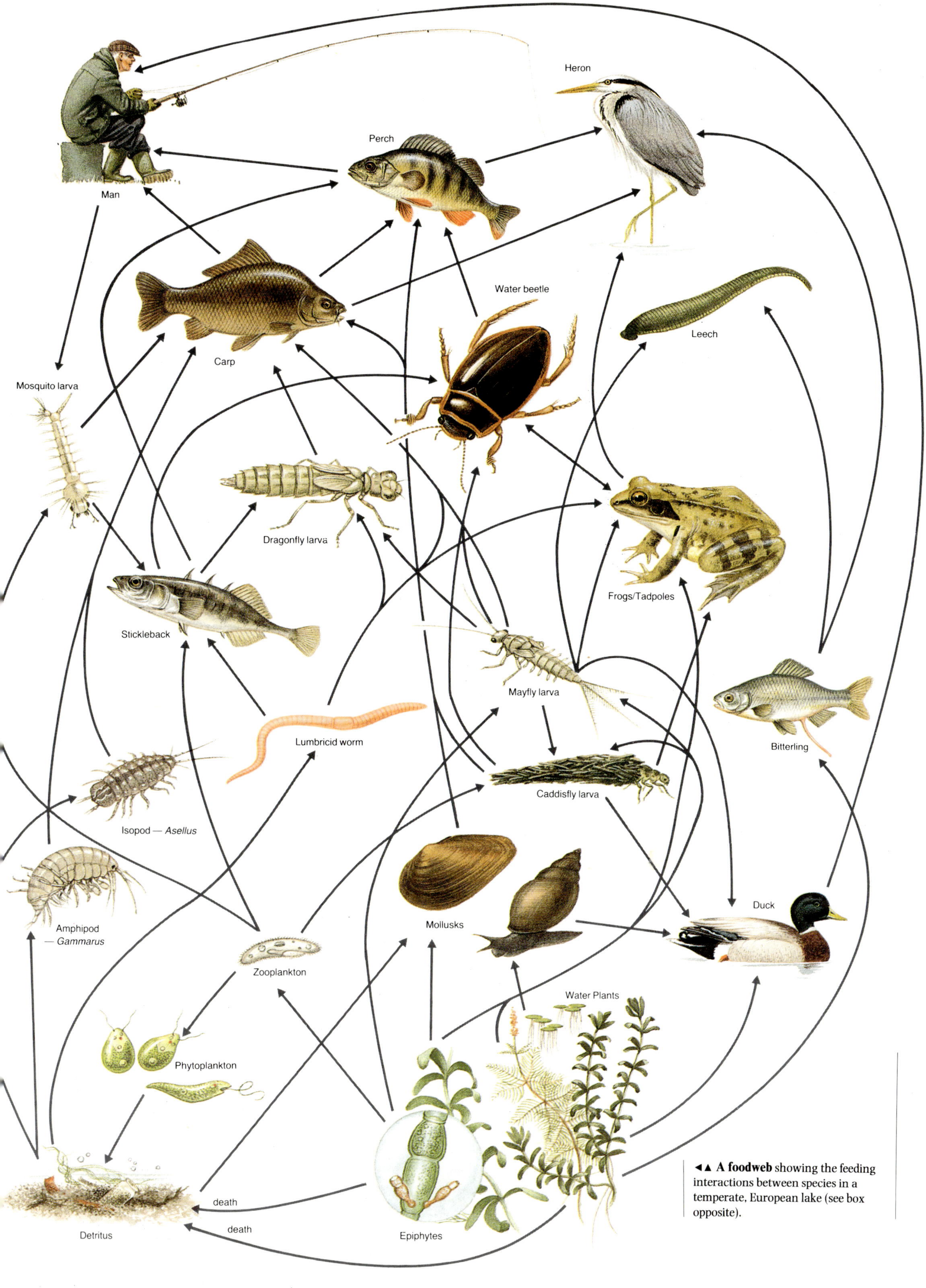

◄▲ **A foodweb** showing the feeding interactions between species in a temperate, European lake (see box opposite).

as the *Rhizobium* bacteria which inhabit the roots of leguminous (pea family) plants, and certain other Cyanobacteria which are associated with fungi in some of the lichens.

These elements, carbon, oxygen and nitrogen, all occur in gaseous form or as gaseous compounds, and this means that there is no barrier to their movement from one ecosystem to another, or even from one part of the world to another. In aquatic ecosystems, they occur in dissolved form. There are some elements, however, such as potassium, calcium and phosphorus, which have no common gaseous form and which therefore cannot move around quite as easily.

Take phosphorus, for example. It is present in certain rock minerals, such as apatite which is weathered in the soil to provide phosphates: these in turn are taken up by plants and hence become available to grazing and then carnivorous animals. In these phosphorus serves a number of important biochemical functions, such as the control of energy relations in cells, and as a constituent of cell membranes; as calcium phosphate it is also an important constituent of bones. Death and decay of the plants and animals releases phosphorus back into the soil, where it may become available again in a local cycle, or it may be carried away by the movement of water through the soil, ultimately being carried to the sea. In the oceans it is used by

How Moose Get their Sodium

An intriguing example of the problems an animal may encounter in its demands for sodium has been brought to light as a result of studies at the Isle Royale National Park, Lake Superior, in Canada. Here, in an isolated and relatively undisturbed ecosystem, moose browse upon the vegetation and their population is preyed upon by wolves. About 1,000 to 1,200 moose occupy the 550 sq km (212 sq mi) area of mixed deciduous and conifer forest, where sodium is not a particularly abundant element. It was calculated that fir trees contain a concentration of about 3 parts per million (ppm), birch 16ppm and aspen 7ppm, which means that the total new browse available for the moose each year as primary production is about 170kg (375lb) for the whole study area. The moose, on the other hand, consume only about 10–20 percent of this available material, that is about 17 to 34kg (35–75lb) of sodium, yet their total sodium output in urine and feces amounts to 243kg (536lb) each year for the whole population. Since there is no evidence of sodium deficiency in the herd there must be some additional source of sodium beside their browse intake.

Licking soil is one way in which moose can supplement the sodium in their diet, but the concentration of sodium in soil was found to be only 24ppm, only about twice that in the browse. It was then noticed that in the summer moose spend much of their time grazing upon aquatic plants while wading in shallow water (see RIGHT). When these water plants were analyzed, it was found that they had much higher sodium concentrations, such as Bottle sedge with 246ppm, Water milfoil with 4,750ppm and bladderwort with over 8,000ppm sodium. Here, evidently, is the moose's source of sodium, but two problems remain unanswered. Why do water plants accumulate such high levels of sodium? And how do moose store the sodium gained in the brief eight-week summer period for the rest of the year?

▶ **Gathering their sodium.** A cow moose (*Alces alces*) and calf feeding on water plants in summer.

The Elemental Components of the Corn Plant and Man

Percentage of Dry Weight

Element	Maize	Man
Oxygen	44.4	14.6
Carbon	43.6	56.0
Hydrogen	6.2	7.5
Nitrogen	1.5	9.3
Silica	1.2	0.005
Potassium	0.9	1.1
Calcium	0.2	4.7
Phosphorus	0.2	3.1
Magnesium	0.2	0.2
Sulfur	0.2	0.8
Chlorine	0.1	0.5
Aluminum	0.1	–
Iron	0.1	0.01
Manganese	0.04	–
Sodium	–	0.5
Zinc	–	0.01
Rubidium	–	0.005

Note the general similarity of the elements required, although man needs more nitrogen (protein) and calcium (for bone structure). One element needed by higher animals and not by plants is sodium, owing to its role in nerve function.

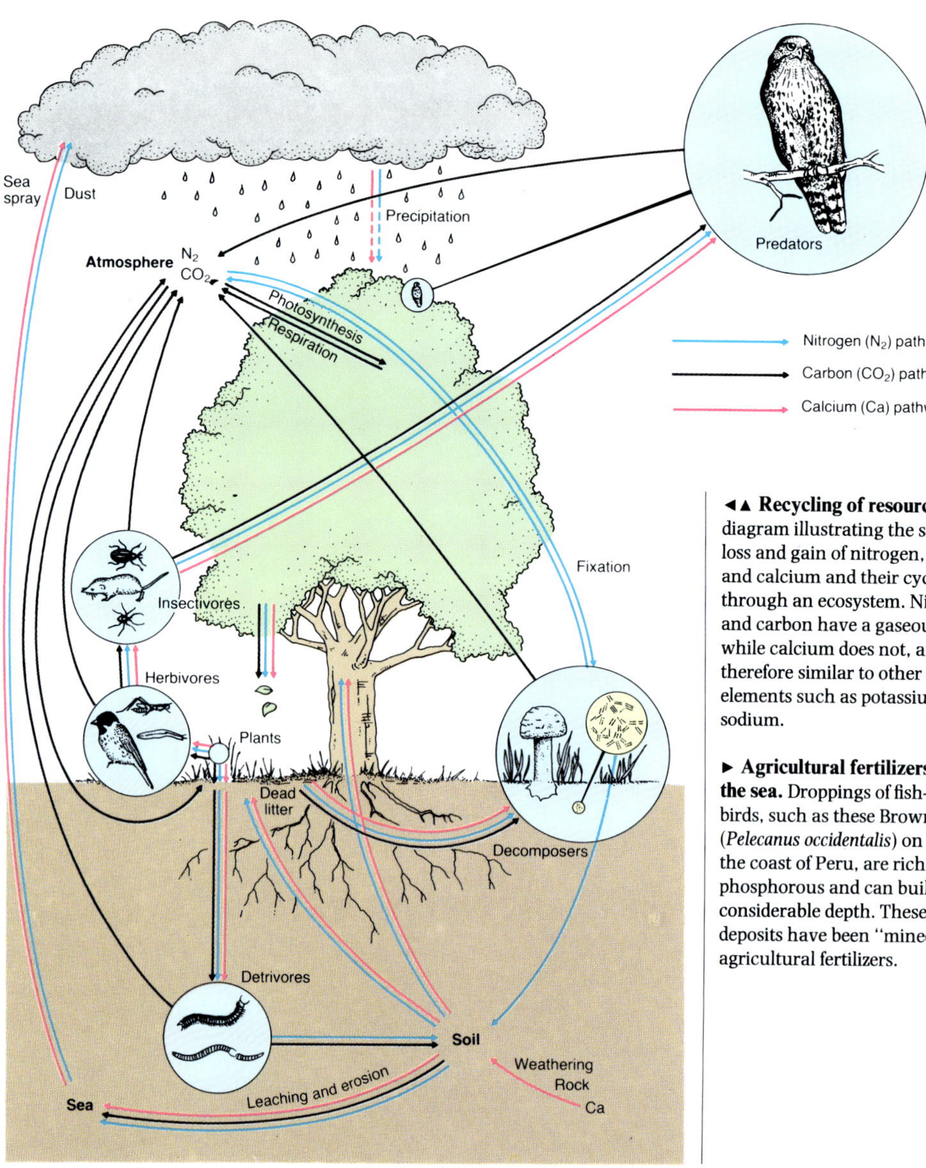

◄▲ **Recycling of resources.** A diagram illustrating the sources of loss and gain of nitrogen, carbon and calcium and their cycling through an ecosystem. Nitrogen and carbon have a gaseous phase while calcium does not, and is therefore similar to other mineral elements such as potassium and sodium.

► **Agricultural fertilizers from the sea.** Droppings of fish-eating birds, such as these Brown pelicans (*Pelecanus occidentalis*) on rocks off the coast of Peru, are rich in phosphorous and can build up to a considerable depth. These "guano" deposits have been "mined" for agricultural fertilizers.

marine plants and animals, and may find its way back to the land when carried by fish-eating birds which leave their droppings on cliff roosts and nesting sites. In the past, these "guano" deposits, especially on the west coast of South America, have been important reserves of phosphorus and have been exploited for agricultural fertilizers. But much of the oceanic phosphorus is deposited on the sea bed in the dead bodies of plankton, and there forms compacted layers over millions of years, ultimately becoming hard rock. Only when the ocean floor is uplifted by tectonic action will these reserves of the mineral be brought back into the cycle once more.

When viewing nutrients on this global cycling scale, it is useful to ask the question whether any of the elements required by living creatures are in short supply and might thereby limit the development of the biosphere. Of the major elements, it could be argued that nitrogen, although not scarce *per se*, is limited in its availability by the microbial (and now the human) fixation of the gas. But it is probably phosphorus which is most widely demanded and yet is by no means abundant in its global distribution. Perhaps this is the most likely element to be limiting the development of the biosphere.

A simple illustration of the way in which nutrient elements can interact and affect the balance of different species in an ecosystem is found in bodies of fresh water polluted by the outwash of fertilizers from agricultural land. The most abundant elements in these fertilizers are nitrogen, phosphorus and potassium, and the fresh waters affected by an excessive input of them are said to be *eutrophicated*.

If both the nitrogen and the phosphorus levels of the pollutant run-off water are in high concentration, then the water bodies receiving it become rich mainly in green algae. Often the whole surface is covered by the green slime produced. The ultimate outcome of this is that the lower layers of water become isolated from the atmosphere and the oxygen dissolved in the water is used up by the animals and the microbes feeding on the dead tissues of green algae falling from above. As the oxygen runs short, so most organisms die.

If the polluting water is rich in phosphorus, but is low in its nitrate content, then growth of the green algae is limited. But, as is usually the case in ecological systems, there is a fierce competitor waiting in the wings. This time it is the free-living, aquatic members of the blue-green bacteria (the Cyanobacteria), which grasp the opportunity presented to them. These primitive organisms are able to fix their own nitrogen from the abundant atmospheric gas, so they are unaffected by its scarcity in dissolved, nitrate form. The blue-green bacteria thus

assume dominance in the water and form a thick scum similar in form to that of the green algal slime, but more blue in color. The ultimate outcome, sadly, is very similar, with oxygen depletion and animal death.

Man has had a considerable impact upon these global cycles in the last century or so. The great changes in the earth's cover of vegetation which have resulted from the clearance of forest and the spread of agriculture, coupled with the large-scale burning of fossil fuels with the expansion of industrialization, have had extensive consequences. The carbon dioxide content of the air, for example, had risen from 314 parts per million (0.0314 percent) in 1960 to over 340 parts per million in 1982. There can be little doubt that this is mainly a consequence of burning fossil fuels—in effect, man is putting back into the atmosphere carbon which has been out of circulation since the coal and oil deposits were laid down many millions of years ago. Man is accelerating a natural cycle. The results of this build-up of carbon dioxide are the subject of considerable debate, but it will probably result in a slower loss of heat to outer space (the "greenhouse effect") that will cause a rise in world temperatures, possibly by as much as 2C° (3.5F°), by the end of the 20th century.

Sulfur is another element which is being discharged into the atmosphere from fossil fuels. Some proteins contain sulfur, and if these are present in quantity in the coal or oil which is burned, then the sulfur enters the atmosphere in an oxidized form (eg sulfur dioxide) and can dissolve in rain water to form sulfurous and sulfuric acids. This "acid rain" is proving particularly serious in areas like Germany, Scandinavia and Canada where soils lack the alkaline lime component to neutralize it.

As well as these global studies of nutrients, it is often valuable to trace the movements of nutrients into and out of ecosystems and to study their storage and movement within the system. Such information is of great value to the environmental manager and conservationist, since it determines whether fertilizers need to be applied to maintain any given level of harvesting from a system such as a forest or a hay meadow.

The study of nutrient cycles on this smaller scale is very difficult for those elements which move freely as gases. It is more easily achieved for elements such as potassium and calcium, which come from outside mainly as dust or suspended in rainfall and which leave in streams draining from the catchment area. Even in the case of these elements, however, there are problems, such as determining the rate at which they are being released into the soil as rocks are weathered. It is also important that the bedrock of the study site should be impervious, so that all the water leaving the ecosystem can be monitored as it passes down a single stream. Other complications, such as migrating animals, roosting birds and leaf litter blown from other ecosystems, can add to the problems of documenting nutrient movements.

Experimental studies of woodland ecosystems in North America, at Hubbard Brook, have provided much interesting information about the way in which nutrients behave. In the case of calcium, for example, 3kg per hectare (2.7lb/acre) arrives each year in rainfall, 5kg/ha (4.5lb/acre) is released from the soil and 8kg/ha (7.1lb/acre) is carried away by the stream. So the inputs balance the output, and the ecosystem is in a state of equilibrium. If the calcium already present in the system is considered, 203kg/ha (181lb/acre) are present in the trees and leaf litter and 365kg/ha (326lb/acre) are in the soil. So the total reservoir of calcium in the woodland is 568kg/ha (507lb/acre), while 8kg/ha (7.1lb/acre) or 1.4 percent of the reservoir is actually passing through each year. Such a system, with a relatively small throughput compared with the reservoir, can be regarded as an essentially stable one.

Such studies also permit an experimental approach. An area can be cleared of all its trees and the consequential disturbance to the nutrient cycles observed. This was done at Hubbard Brook and the results proved dramatic. Water runoff increased by 40 percent, because the trees were no longer intercepting rainfall and extracting water from the soil. The output of nutrients was impressive: calcium discharge increased four times, potassium 15 times and nitrate by about 50 times. The great reservoir of elements in the trees and the litter was being released back into the environment.

A knowledge of the nutrient reservoirs and movements within ecosystems, therefore, helps in the understanding of how those systems are balanced and how they function. And this provides the clue to their management and conservation.

PDM/BDT

SPECIES INTERACTIONS

Why populations change. . . Competition for limited resources. . . Population size depends on births, deaths, immigration and emigration. . . Species that depend on each other. . . Species that prey on other species. . . Ecological niches. . . Communities of different species. . .

WITH animals, as with human beings, relationships are important. No individual is isolated from others of its species—and no species lives wholly unaffected by the other species around it.

Population is the term used for a collection of individuals of the same species which live in the same area at a given time. It is in fact a very flexible term. The boundaries of a population are set by whoever is making a study of it, so that it is just as correct to talk about the population of aphids on a single wheat plant as it is to consider the aphid population of an entire wheat field. But there is an important point: to understand the way in which a population functions, population boundaries must be identified in the same way that individuals of the population see them.

Populations change in two main ways: in space and in time. Returning to the example of a wheat field, the populations of aphids on separate wheat plants will differ throughout the field as a whole. Some plants will have no aphids, while others will be heavily infested. Moreover, the wheat plants in some parts of the field will have many aphids while those in other parts will have few. This patchiness is a common feature of most natural populations. Looking at the aphids at different times of the year, their population size changes through time, becoming sometimes bigger, sometimes smaller. So the numbers of individuals in a population vary through time and space. Populations also differ in other ways. The ratio of young to old varies, as does the genetic makeup of individuals.

These differences in time and space are usually the result of a natural patchiness of resources in the environment, together with interactions with other individuals and species.

In an unlimited environment all species have the potential to multiply at an increasingly fast rate. In the natural world, however, environments are not unlimited. As the population gets bigger and bigger, so one or more of the resources needed by the population—such as food, water, nutrients or space—becomes scarce. This gradual decrease in available resources slows down the rate of population growth, either by reducing the number of births or increasing the number of deaths, or a combination of both. Eventually there will be a balance between births and deaths such that the population can just be supported by the available resources. Exactly which resource is the limiting factor, and at what level the population stops growing, depends on local conditions. For example, in the wheat field some wheat plants will be growing more vigorously than others, perhaps because the soil is moister or richer in nutrients, and so the aphids will do better on these plants than on the others. Predators, such as ladybugs and lacewing larvae, will also concentrate in those areas of the field where the aphids are growing more vigorously.

In this example the habitat can be seen to influence the size and structure of population. An even more evident influence of the habitat is the occurrence of a catastrophe that causes a sudden decrease in a population. Part of the "patchiness" in nature may be a consequence of such unpredictable events.

Four processes are involved in actually causing the changes in population size. These are births, deaths, immigration and emigration. Births and immigration increase the size of the population. In a favorable habitat young can be reared successfully, and the good conditions will attract others to the area. Deaths and emigration lower the numbers in a population. In less favorable habitats, the numbers dying will increase, and surviving individuals will try to escape to a better place.

When a species reproduces, its young have to grow and develop before they themselves can reproduce to perpetuate the species. These youngsters face many hazards before sexual maturity, and there are two main methods of reproduction that

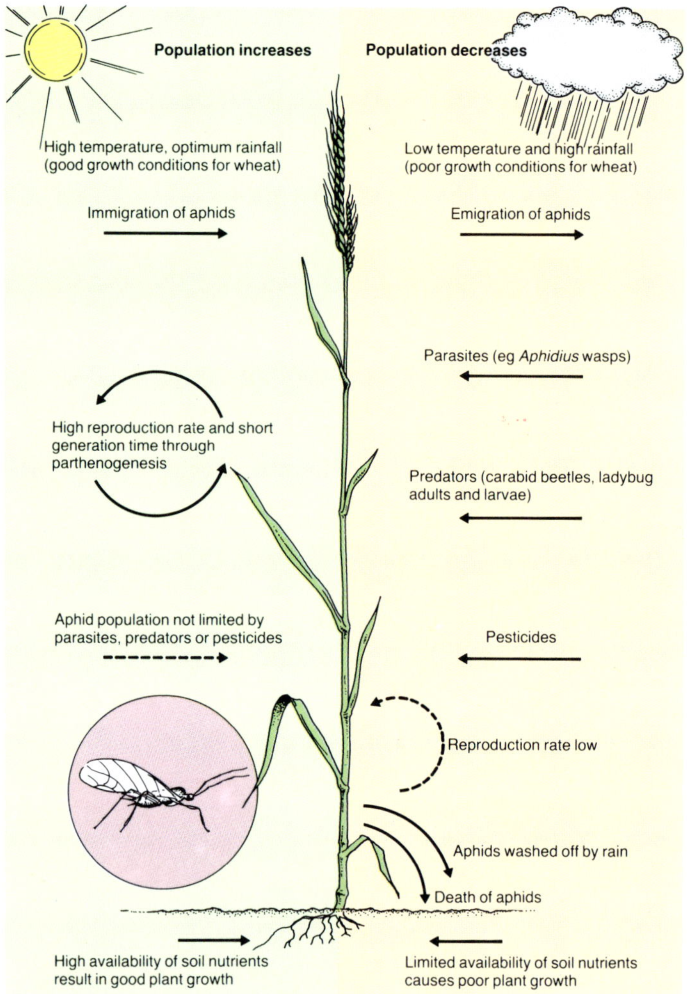

▲ **Population variation**—factors affecting changes of population size, as exemplified by aphids that feed in wheat fields (see text for detailed explanation).

► **Crowned crab**—a commensal relationship between a hermit crab and sponge in the water off the Cape coast of South Africa. The crab obtains camouflage by this association, but the sponge apparently gains nothing, although it may derive some nutrients from scraps drifting to it while the crab feeds.

ensure that enough reach this stage. In some species, for example oysters and dandelions, the chance game of survival is played with very large numbers of eggs or seeds which are left to fend for themselves. Most of these will fail to survive. At the other extreme each female produces only one or two offspring at a time, which with considerable parental care have a good chance of reaching maturity. Examples of this pattern are seen in penguins, monkeys, humans and Tsetse flies.

So far each species has been considered in isolation. This is frequently done in laboratories, and can yield an understanding of several important population properties. In the real world, however, no species are isolated. They interact in various ways with other species in the same habitat or area. Some of these interactions are beneficial, while others are not.

Species referred to as "symbiotic" are those that live together for their mutual benefit—often, indeed, being unable to survive on their own. Lichens are well-known examples, where an alga and a fungus form an intimate relationship. A less unified symbiosis is seen in the several instances of cleaners. Oxpeckers in Africa feed on external parasites of large mammals, to the obvious advantage of both, and some reef fish (wrasses) carry out a similar service for their neighbors.

In some cases the relationship is one-sided (*commensalism*), with one species clearly gaining from the situation while the other derives nothing or perhaps loses out. The many epiphytic plants growing on trees, using them only as a support or substrate, provide an example of this type of commensalistic relationship.

Among animals, sea anemones living on the shell of a hermit crab are another example of commensalism. The crab gains little but the anemone is provided with substrate and a ready food supply in the form of scraps from the hermit crab's meals.

Parasites, predators and herbivores all gain from the species they feed on. In the case of predators their prey lose totally and are killed, but parasites and herbivores usually take only a part of the individuals they feed on and leave them to survive even though in a weakened state. In much the same way that a complex symbiotic relationship evidently needs a long evolution to bring it to its present fine-tuned state, so parasites, herbivores and predators have all gradually evolved alongside the species they exploit.

The population of a single species is prevented from increasing indefinitely by mechanisms that increase the death rate as the population gets larger. There are several types of mechanism that bring this about. One of the most direct and important is shortage of food. Predators feed on their prey. If the pressure on the prey is too great, then there will be insufficient food for the predators and their numbers will decline through starvation. Some studies of this interaction between predator and prey suggest that it is oscillatory; populations go up and down alternately. Predator numbers decrease owing to starvation; this releases the remaining prey from strong predation pressure, and so their populations can increase. As the prey become more abundant, so the predators can more easily get their food, and they also begin to increase. Both populations continue to grow until the predators again begin to take too many prey causing their population to decline once more, and the cycle begins again. The extent and frequency of these oscillations is governed by differences in the rates of reproduction of predators and prey, and also by whether the predators have any other prey species on which to feed. Some ecologists, however, feel that this is too simple a view of predator/prey interactions.

Herbivores are sometimes limited by their food, for example the White rhinoceros and other large grazing herbivores of the African savanna, though most herbivorous insects are not limited by food availability but controlled by their predators. Parasites and disease are also important in controlling populations. At high population densities diseases spread rapidly, and if fatal they cause the population to plummet. Non-fatal diseases weaken affected individuals and have more subtle effects on the reproductive rate: for example, mumps can cause sterility in male humans. Parasites also tend to debilitate and reduce the vigor of their hosts. Several insect groups including bees, wasps and plant bugs are parasitized by another insect group called stylops. Females attacked by stylops change in form into males and are referred to as being "stylopized."

In some species, such as voles, it has been suggested that

► **Crowded nesting colony** of King penguins (*Aptenodytes patagonicus*) on Crozet Island in the Subantarctic—the birds are holding the single egg on their feet below the fold of belly skin. Species such as these produce just single young, with parents investing much effort in raising them; in such circumstances the chances of young reaching adulthood are high. The high density results in a remarkably regular spacing of individuals.

their populations have an inbuilt mechanism whereby at high densities crowding affects their physiology and reproductive behavior, causing a drop in the birth rate.

A species lives where it does because that particular place provides all the necessary resources for its continued success. These resources include physical conditions, such as shelter, temperature, rainfall and light, as well as food and nutrients. In addition, potential predators, parasites and competitors are not present in large enough numbers to prevent the species persisting in that place. The unique occurrence of the correct combination of physical and biological factors, which are characteristic of and different for each species, is called the *ecological niche*. This concept is not to be thought of as just another name for a species' habitat; it also involves the interrelationships between the species in the habitat. Charles Elton, an eminent English ecologist, suggested that the term *niche*

The Battle between Animals and Plants

Through evolutionary time plants and the animals that feed on them have been engaged in combat. Individuals differ one from another. One plant may, because of some slight difference in its metabolism, be less attractive to herbivores than others of the same plant species. This individual will produce offspring which are also less attractive. If the herbivores have a marked effect on the attractive individuals, so that they suffer and produce fewer offspring, then over a period of time there will be a gradual increase in the number of the less attractive individuals in that population. Natural selection ensures that the useful characteristics of a population are conserved while the poor features are lost.

While plant populations become less palatable owing to the natural selection of the herbivores, the herbivores themselves are also changing. Slight differences allowing an individual herbivore to deal with the unattractive qualities of a plant will allow that animal to have the plant to itself. So there is a continuing battle being fought between plants and their consumers. The plant weapons in this battle are primarily chemical, although some resort to spines or hairs to reduce their attractiveness to herbivores.

A few plants, such as acacias (ABOVE) and *Barteria*, use biological weapons. They have developed symbiotic relationships with ants. In return for the protection the ants offer in keeping the plant clear of herbivores, it provides them with food from nectaries on parts other than flowers and suitable nest areas in the form of hollow stems or, as in the acacia, swollen bases to its spines. The ants not only keep the plants free of insect herbivores, they are quite unpleasant enough to drive off large mammals, including man. In another plant/ant relationship between the tree *Cecropia* and the ant *Azteca*, the ants also remove potentially choking vines and creepers from their host.

▲ **Home among the thorns**—*Crematogaster* ants nesting in specially-adapted *Acacia* thorns, Kenya.

◄▲▼ **Different professions** in the same community—dividing up the desert. This sequence of photos illustrates some of the niches for small mammals in the Kalahari. ABOVE A rock mouse (*Petromyscus* species), which is common around human habitations. LEFT Cape ground squirrel (*Xerus inauris*), which digs for roots and tubers, and plucks grasses at their bases. BELOW LEFT Bushveld elephant-shrew (*Elephantulus intufi*), which specializes in eating termites, the most abundant animal food resource. BELOW South African pygmy gerbil (*Gerbillurus paeba*), which is commonest on the dunes, feeding upon and hoarding seeds.

includes a consideration of the "profession," as well as the "address," of a species.

The ecological niche can be considered on two levels, with and without competitors. Imagine a village with two shops. If both shops stocked the same items they would be in direct competition and eventually one or even both would go out of business. In ecological terms the two shops would have had identical niches. The outcome might be otherwise if the shops stocked different items. There is now the possibility that both would succeed. They are very similar in their habitat, but they differ in their professions (a grocer and a butcher, perhaps). A similar situation occurs in nature. Species with identical niches cannot live in the same place together. However, small but important differences in their niches will allow coexistence, because the species are partitioning, or dividing, the available resources. Much of what is called the evolution of new species is the result of individuals gradually changing profession or habitat under the influence of competition.

The populations of different species living together in any one place form a community. The niches of these species overlap, in that they all experience the physical characteristics of the habitat, but they all differ slightly in their professions. A community will have a range of carnivores but these will each feed on prey of differing sizes. In the same way the herbivores will feed on different plants in the area or on different parts of the same plant. Limiting resources may not be food but some other requirement. Plants need light, and nutrients. Other species, normally excluded by competition, may even temporarily become part of a community if some necessary resource is abundant. For a short time potential competitors can coexist in the same way that a traveling salesman may for a short period make a living in the village. These are transitory species, and are often adapted to a roving existence among established communities by having good powers of dispersal. "Weeds" exist like this.

Species in a community are held together by the foodweb of feeding interactions. Species of any one trophic level compete with others at the same level for the energy available from the trophic levels below them. This view of a community as a collection of species living and interrelating in the same place is zoological in origin and based on function. It stems from Elton's idea that feeding links are all-important.

Botanical ecologists, by contrast, have tended to think of communities in a different light by concentrating only on the plants. They noticed, early in the development of ecology, that collections of plant species tend to occur together. These collections were called *associations*. Confusion developed because the term "community" was often used instead of "association"—people talked of the "oak-wood community" or the "old field community" when thinking of the collections of plants that characterized these habitats.

While there is no doubt that groups of plant species *do* associate together, such ideas are far less clear-cut if animals are considered as well. Indeed, it was Elton once more who pointed out that in different habitats there were "ecological equivalents," different species with the same profession; for example, in the New World hummingbirds feed on nectar, but in Africa it is the sunbirds that do so. BDT

PHYSICAL CONSTRAINTS

Life-forms limited by temperature and water availability... How animals and plants specialize... The effect of climate on plant distribution... Plants can create local conditions... Limitations to animal development... How land animals adapt...

LIFE on earth depends principally on two things, because of the nature of living tissues. One is the availability of water in one form or another, since the cell contents (protoplasm) consist largely of water (85–90 percent). The other requirement is a suitable temperature, since the proteins of which animal tissue largely consists are destroyed by extremes of temperature. Thus where no water is present, even periodically, life cannot exist except for a short time. Where temperatures are continuously below −10°C (14°F) or above 45°C (113°F) life is also necessarily absent. Fortunately a very high proportion of the earth's surface fulfills these conditions for life.

From the beginning, animals and plants have been subjected to evolutionary pressures, particularly competition for resources from other species and groups. These pressures led to plants and animals becoming specialized for various ways of life and habitats, by acquiring characteristics which enable them to function most effectively in these circumstances.

This struggle for resources and consequent specialization led species to adapt to increasingly stressful environments—hence the original invasion of relatively inhospitable freshwater and land environments from the sea.

As a result of this evolutionary process, all species have their physical limits which define their degree of specialization. The chief limiting factors are temperature and tolerance of water loss, but there is a whole series of others including light, the concentration of various inorganic ions such as calcium, the acidity of the soil, and salinity. Thus, for example, few earthworms occur in acid woodland soils, and few freshwater fish can tolerate the salty water of estuaries. The distribution of a species will occur only within its limits of tolerance of these factors. The species will be most successful where conditions are most suitable for it, and will be less successful where it is approaching the limits of its tolerances. However, it would only be in an environment devoid of all other competing species that a species could extend to its tolerance limits. Such a situation rarely exists in nature, although this might occur on a newly formed and isolated island. It is important to recognize that for plants and sedentary animals it may be the tolerance of the young and not of the adult which sets the limits of distribution. So although oyster larvae and seaweed spores may settle above adults on the shore, they are rapidly killed by drying and also perhaps by high temperatures. Thus study of tolerance limits for the whole life history may be necessary to understand the observed distribution. For example, low winter temperatures,

► **The hospitable environment of the sea**—Green turtle (*Chelonia mydas*) and fish. The physical conditions within the sea remain fairly constant, but the main problem for marine animals is preventing excess water loss through their body walls, as their body fluids may contain less dissolved substances than the seawater.

especially frost, kill seedlings and restrict the northward spread of Loblolly pine in the USA. Elsewhere, the distribution of an organism will be restricted to areas where the species is able to compete better than all other species with similar abilities. Clearly, where the species has optimum conditions, it is more likely to be competitive and successful. Good illustrations of this principle in action can be seen where physical conditions change rapidly from one place to another nearby, such as at the junctions between air and water.

Animal Distribution

For animals, the importance of physical factors in determining their distribution depends on the environment they inhabit. The seas, where most animal groups originated, are the least stressful environment. Marine animals are thus highly evolved and can be expected to be well adapted to their environment. Although there is a considerable range of temperatures from the tropics to the poles, nowhere is animal or plant life precluded. Nevertheless, temperature is important in determining the distribution of marine species. Some species appear to have a distribution entirely controlled by temperature; thus, for example, most corals are restricted to water with a temperature not falling below 18°C (65°F). Most species have an optimum temperature for their functioning and become less competitive near their limits. For many species a suitable temperature must be reached for reproduction to take place. Populations near their distribution limits are often effectively sterile and incapable of spreading further because this temperature is not achieved. Such populations are either ephemeral or replenished by recruits from warmer water. This may account, for example, for the restriction of a number of warm-water intertidal species to the southwest of the British Isles.

In the intertidal zone, the boundary zone between the sea and the land, the effects of physical constraints on plant and animal distribution are dramatically displayed. At low tide, the plants and animals of the marine intertidal are all subject to desiccation, temperature extremes and exposure to ultraviolet light, with the extent of these stresses increasing up the shore. The upper extent of species living here is determined by their tolerance of such conditions. The death of larvae or sporelings which settle above the correct zone demonstrates that conditions above the zone are intolerable. In fact most animals and plants of this habitat have a relatively narrow band or zone in which they are successful. This reflects the fact that it is only within this band that they are competitively superior to species living above or below them.

Evolution has resulted in clear-cut adaptations for resisting the physical hardships of this intertidal environment. Among the algae these include thick cell walls with mucilage coats, resulting in a considerable tolerance of dehydration. The latter capacity is shared by the animals, many having shells which resist water-loss and which may be pale so that they reflect excessive solar heat. Animals have also evolved the ability to produce waste products containing little water, and many are capable of breathing in air. Some, like winkles, talitrid amphipods and some tropical crabs, are virtually terrestrial animals. Despite these adaptations, many forms, such as limpets, live near their limits and they therefore have to leave exposed hot surfaces for cooler, more shaded sites in hot summer weather.

In the terrestrial environment all animals must be able to tolerate dry conditions and to function in a variable temperature regime—or they must have some means of avoiding these extremes. Mobility is essential to escape from unusual and dangerous situations and thus there are few sedentary terrestrial animals. In dry conditions all animals lose water through their urine and feces, their respiratory surfaces, since these must be kept moist, and (in ill-adapted species) through the general body surfaces. Such water loss depends on temperature. The success of a species is measured in terms of its ability to restrict this water loss or to tolerate water loss, and the extent of its distribution will also thus depend upon these factors. By these criteria, few groups of animals are truly successful in the terrestrial environment. Soft-bodied creatures such as worms, slugs and crustaceans exist in damp areas and are joined there by vertebrates with little adaptation for water retention, such as amphibians.

Well-adapted creatures among the invertebrates include insects, spiders and scorpions, which have a waterproof skin (*cuticle*), and among the vertebrates reptiles, birds and mammals.

Every garden in the temperate zone demonstrates that some species are limited by dry conditions. Woodlice, slugs, centipedes and amphibians will be found by day only in humid, cool,

▼► **Zonation and animals in the intertidal zone** of a rocky shore in southern Africa. In the littoral fringe or splash zone, which is rarely covered by water: (1) a sea slater *Ligia natalensis*. On the upper eulittoral, which is only submerged at high tide: (2) a periwinkle *Littorina africana*; and acorn barnacles (3) *Chthamalus dentatus* and (4) *Tetraclita squamosa*. On the mid eulittoral, which is exposed to the air for about 12 hours in each 24 hours: (5) a periwinkle *Oxystele sinensis*; (6) Brown mussel (*Mytilus perna*); and limpets (7) *Patella cochlear* and (8) *Patella barbara*. On the lower eulittoral, which is covered by the sea for most of each day: (9) a tube worm *Pomatoleios crosslandi*. In the sub-littoral fringe which is only very rarely exposed to the air: (10) a soft coral, dead-mans-fingers (*Alcyonium fallax*); and (11) a sea squirt *Pyura stolonifera*.

Tolerance Limits of the House Fly

The House fly is a familiar creature which has its center of population in warm climates. In temperate areas like Europe and the northern USA it is active only in summer, for reasons illustrated below:

Death occurs in 40 minutes at −5°C (23°F)

Enters chill coma at 6.0°C (42.8°F)

Just able to move at 6.7°C (44.1°F)

Feeble movement possible at 10.8°C (51.4°F)

Normal activity at 15–23°C (59–73°F)

Rapid movement at 28–35°C (82–95°F)

Excess activity at 40°C (104°F)

Heat coma at 44.6°C (112.3°F)

Upper lethal temperature 46.5°C (115.7°F)

► **Sheltering from shade** OPPOSITE, a pair of Arabian oryx (*Oryx leucoryx*). To help prevent evaporative water loss, this species can allow its body temperature to increase during the hottest parts of the day.

► **Unlikely inhabitant of deserts**—a Trilling frog (*Neobatrachus centralis*) emerging from sand after rain. Several species of frog inhabit deserts. They lay eggs in temporary pools after the infrequent rains. The full development cycle from hatching through tadpoles to mature frog must take place in a matter of weeks. As the pools dry up the frogs dig deep into the sand and there remain dormant, sometimes for years, until the next rains.

dark refuges, such as under rocks, logs and leaves, whence they will emerge at night when it is both cooler and damper. Their behavior patterns ensure that they remain in suitable conditions. Such gardens also contain invertebrate and vertebrate species well adapted for terrestrial life. Their distribution patterns are less obviously linked with physical conditions. However, all have tolerance limits, as shown, for example, by the common House fly (see table).

Competition for resources in the terrestrial environment has led to animals extending into more and more demanding situations, with corresponding physiological adaptations and specialization. As conditions become more extreme, fewer groups can survive, and the variety of species in a desert is very restricted compared with those in a damp rain forest. However, species do exist—sometimes surprising ones are found—and the final determinant of whether an animal is present in a desert is often the availability of food rather than physical conditions. Snails and toads would seem unlikely inhabitants of deserts, but the snail *Sphincterocheila* lives in semi-desert (see below).

Some small mammals can exist where no drinking water is available, obtaining water from their food alone and surviving because their physiology and behavior is such that their water losses and therefore requirements are minimal. Kangaroo rats are a good example. They are seed-eating, burrowing, nocturnal mammals. Their water loss from the feces is very low, their urine is four times as concentrated as that of humans, and they have no sweat glands. Most water loss is thus from respiration. They are active at night, spending the day in their small burrows beneath the soil, where the temperature is much lower than at the surface and where the exhaled air remains humid, thus limiting water loss.

A Desert Snail

Snails are unlikely residents of deserts, but one (*Sphincterocheila*) lives in semi-desert conditions where its lethal limits for temperature are often exceeded and water is available only at infrequent and unpredictable intervals. Yet it survives in surprising numbers, owing to its adaptations.

The snail's lethal limit for temperature is around 55°C (130°F), but surface temperatures regularly reach 65°C (150°F). Why does it not die? Firstly, the shell is white and so reflects the sun's energy. It is also helped by its shape, so that 95 percent reflection is achieved. Heat is not absorbed from the ground, because the animal's shell touches the ground only at a few points, and, except when active, the animal retracts into the shell leaving an insulating air layer. Thus the temperature of the snail tissue is only 50°C (122°F) when the ground is at 65°C (150°F).

The shell also enables the snail to retain water. It is thick, has a small aperture, and the mucous membrane, which can be thrown across the aperture to impede water loss, is also robust. The snail can survive more than three years' dehydration—and then, when rain occurs, it rapidly becomes active and takes advantage of the suitable conditions.

Similarly, toads can bury themselves for long periods and enter virtual suspended animation, only to revive remarkably quickly when water becomes available.

Plant Distribution

On a global scale there are clear gradients in mean temperature from the equator both northward and southward. The greatest temperature range is found at the centers of large Northern-Hemisphere land masses, far from the moderating influence of oceanic water. Clearly, a plant species will be found only in those parts of the earth where the temperature range is within its tolerance limits. Rainfall is another, related, factor in determining plant distribution, the two together forming the basis of climate.

For the plants themselves, it is the actual availability of water that is crucial. Therefore, the rate of evaporation is as important as the rainfall, since most precipitation moves directly back to the air through evaporation and only a small proportion enters streams and rivers as runoff water. Evaporation depends on the temperature, and thus low rainfall and high temperatures are more likely to give extremely arid conditions than low rainfall and low temperatures.

The presence of mountains, and even small variations in topography, influence the climate and hence the vegetation patterns. Temperature and rainfall patterns alter drastically as major mountains are climbed, and this results in vegetation bands resembling latitudinal ones. Often one side of a mountain range may receive less rain (rain shadow) and thus have an entirely different flora from the other side. High mountain ranges may act as barriers in this way to the geographical spread of species unable to tolerate conditions at high altitudes, and valleys within high mountains may be isolated for so long that evolution to the local conditions causes the plant forms within the valley to become a distinct entity—speciation through isolation.

Correlation between vegetation type and climate is normally so good that geographers may classify the climate on the basis of the vegetation, and botanists who study fossils can tell what the climate of an area was like thousands of years ago simply by determining what plants were present at the time.

Major vegetation patterns are thus determined by plants' tolerances of physical conditions and also by soil characteristics such as acidity or alkalinity, calcium content and nutrient availability. However, on a local scale, the plants themselves may be responsible for modifying local conditions, creating climatic differences on a small scale. Under any land plant there will be shade, and in that shade the temperature and light intensity are lower, and the humidity higher. The extent of the change will depend on the plant size, foliage density and other factors, but the potential importance of less extreme conditions for seedlings, tender plants and vulnerable animals will be obvious. Many young plants can only survive in the shade.

Of course, plants are not the only cause of micro-environmental differences. Any object creating shade or shelter, such as walls running east–west or seashore rocks with north- and south-facing sides, will have shady and sunny faces and therefore very different microclimates on each side. Each microclimate may have different plants and animals associated with it. Plants or animals near the limits of their tolerances may exist in isolated pockets of acceptable conditions in an otherwise hostile environment as a result of microclimatic differences.

RHE

ECOSYSTEM DEVELOPMENT

How a new ecosystem develops... Climatic and geographical influences... The development of sand dunes as an ecosystem... The development of scrub and woodland... Re-establishment of an older ecosystem... Limits of an ecosystem... Nutrient cycles... Complex ecosystems: are they more or less stable?...

How does a living community develop when an area of ground is made available for colonization by plants and animals? When, for instance, a glacier retreats leaving an expanse of crumbled rock and detritus, or sand builds up around a shore line as it is dried and moved by the wind, a process of gradual development is initiated which will ultimately lead to mature, stable vegetation. The outcome of this process, which ecologists call *succession*, is predictable. Normally, the development continues until as much vegetation is formed as is possible within the limits of the resources of the particular site. Often this means that the characteristic biome for the particular climatic zone will be attained; this is certainly the case if climate is the major limiting resource. There are occasions, however, when other factors limit the degree to which an ecosystem can develop. For example, a site which lies within the climatic region normally characterized by deciduous forest may fail to achieve complete tree cover because the soil is shallow, fast-draining, or nutrient-deficient. This may result in scrub vegetation being the tallest form attainable.

On the other hand, there are some situations in which local conditions may permit the development of an ecosystem which has a greater *biomass* (a greater development of vegetation) than that normally associated with that climatic zone. For example, in the temperate grasslands—that is, the steppes and prairies—the climatic factor which limits growth is summer

drought. In particularly wet areas, however, for example beside rivers, or in water-holding hollows, woodland may develop in response to the additional summer water supply. So the final ecosystem type at which the succession comes into a state of stability or equilibrium may have a biomass somewhat more than or less than that of the characteristic biome type. But if the physical constraints and resource limits of a site are fully known, then the outcome of a succession can be predicted.

Successions can be divided into two main types. There are those which result from a piece of land coming into existence which has not formerly borne a vegetation cover, such as the new land formed by volcanic lava flow, or a newly-formed estuarine mud bank. The succession which occurs under these circumstances is termed *primary*. The other type, *secondary* succession, results from the destruction of an ecosystem, either by natural or human agencies, and its subsequent recovery.

Both of these succession types lead towards an increased biomass, and they may eventually produce a similar stable end point, or "climax," if all other factors are equal, but the starting point is quite different. This means that the speed and even the course of the succession may differ.

One of the best examples of a primary succession is the sand dune. Sand from the intertidal zone is dried by the sun at low tide. Some is then blown by the wind, accumulating above the

◄▼ **Animals of Coto Doñana** sand dunes, Spain. In mature Umbrella pine (D): (1) Spanish Imperial eagle (*Aquila heliaca adalberti*); (2) Azure-winged magpie (*Cyanopica cyana*). In invading Umbrella pine (C): (3) Short-toed eagle (*Circaetus gallicus*); (4) Red deer (*Cervus elephas*); (5) Montpellier snake (*Malpolon monspessulanus*); (6) Great gray shrike (*Lanius excubitor*). In grass and sedge (B): (7) Crested lark (*Galerida cristata*); (8) Wild boar (*Sus scrofa*). Zone (A) is the advancing dune ridge. See box right for further details.

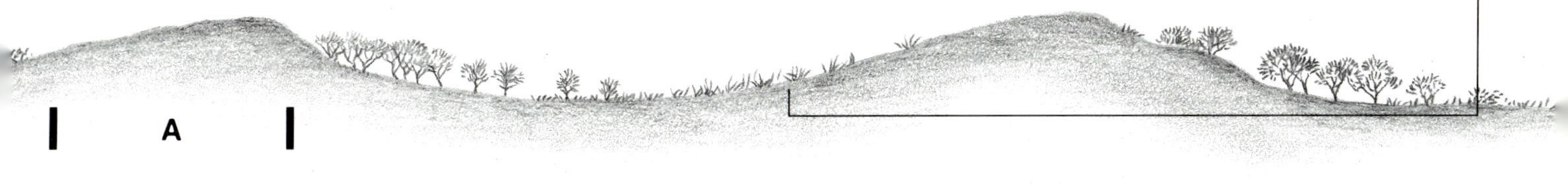

Shifting Sands in Spain

Some ecosystems experience a repeated and often predictable disturbance or catastrophe, such as fire or wind blow in forests, and this initiates the succession anew. One particularly striking example of this is known to occur in the dune systems of southwest Spain, in the Coto Doñana National Park. Here the dune ridges are never fully stabilized, and they move in advancing waves from the Atlantic coast inland, engulfing the developing vegetation. As each dune ridge advances (A), it leaves behind a bare area of flat, damp sand, which is rapidly invaded by grasses and sedges (B). The major tree species of the area is the Umbrella pine, and this soon arrives also (C), taking advantage of the buildup of organic matter in the damp soil. Its winged fruits are easily carried by the wind, but they are also an important food source for animals and birds, particularly the Azure-winged magpie (2), which haunts the local pine woodlands, and these birds undoubtedly assist in the spread of the fruits.

Within about 40 years, the pine trees have formed a mature strip of forest (D), which is producing cones of its own as well as supporting a bird life which includes the magpie. The forest also provides nesting sites for predatory birds such as the Short-toed eagle (3) and the very rare Spanish race of the Imperial eagle (1). But it is at about this stage that the predictable disaster strikes and the woodlands are buried once more by the next wave of mobile dunes.

The Umbrella pine is very well attuned to this sequence of events since it can complete its life-cycle within the allotted 40 years or so. Slower-maturing species such as juniper and Cork oak fail to maintain a viable population under these conditions. There is also evidence that this sequence has persisted in the area for a long time; fossil peat deposits buried beneath the sands and exposed in neighboring cliffs provide evidence from the fossil pollen grains preserved within them that the same cyclic succession has been going on for at least 13,000 years.

strand line. This drifting sand is an extremely difficult substrate to colonize, for a number of reasons. It consists of quite large grains of silica, so it does not weather to produce nutrients available to plants; it contains very little organic matter, so it drains freely and is subject to drought; there are very few microbes, so there is little recycling of nutrients; it is saline and may still be subject to occasional flooding by the sea; the daily range of temperature is very large, since the days are hot and the nights are cold; the sand is always on the move, so it may erode from around the plant, or pile on top of it and bury it. There are very few plants which can cope with such adverse conditions and, in the absence of primary production, animals and microbes are scarce.

One plant which can cope is the grass called the Sand couch. Although it grows to a height of only about 60cm (2ft), it extends by underground stems and binds the sand together. Its upright leaves are drought-resistant and are not desiccated by drying winds. Indeed, the very presence of the leaves causes eddies in the surface airflow, and this causes the driven sand suspended in the air to be deposited in the lee of the obstruction. Small dunes up to 1m (3ft) in height may develop as a result, thus raising the surface of the sand out of the reach of the tidal flow. This then permits the invasion of species of plants which are more sensitive to salinity. Among these is the Marram grass, a more robust and aggressive species than the Sand couch, which can grow to about 1.2m (4ft) in height and which has much more extensive rhizomes. Its dense growth has a profound effect upon the wind speed over the dune surface, and therefore upon the deposition of sand. Under the influence of the Marram grass the dunes grow very quickly and may attain heights of 100m (330ft) or more.

But the dense growth of the Marram grass has a further effect: it suppresses the growth of the Sand couch, which then disappears from the growing dune. The situation is rather ironic, for the Sand couch by its very growth and success in the habitat has created the conditions in which the newly-arrived species has an advantage over it. Merely by growing there the Sand couch has effectively signed its own death warrant!

But the story does not end there, for the continued growth of Marram grass and the buildup of dunes leads to further changes in the microclimate and wind-flow patterns of the entire habitat. While the dune is actively growing and unstable, the Marram grass grows vigorously and flowers profusely, but meanwhile the Sand couch will have invaded new territory in front of the growing dune and will be engaged in the reclamation of new land from the sea. The growth of the new ridge protects the old one, and the sedimentation of sand is reduced. As the dune stabilizes, the growth of the Marram grass slows down and the plants become moribund and weak; they soon give way to other plants. If the soil is fairly well supplied with calcium, shrubs such as Sea buckthorn and privet may flourish; otherwise pine species may assume dominance. The scrub itself may persist or give way to forest if the water table allows forest trees to become established.

In this type of succession, one of the most obvious features is the way in which the growth of the plants themselves modifies the habitat and thereby permits the entry of less specialized, but often more competitive, species. Organic matter is added to the soil, which gives it greater water-holding capacity and encourages the growth of microbes. These provide a faster and more efficient cycling of essential nutrient elements, which

▼▶ **Succession on a sand dune**—animals that colonized a sand dune at each stage of its development. (A) Drift line: (1) Sand hopper (*Gammarus locusta*). (B) Pioneer dune: (2) Common tern (*Sterna hirundo*), nesting site. (C) Mature dune; (3) Oystercatcher (*Haematopus ostralegus*), nesting site; (4) Banded hedge snail (*Cepaea nemoralis*); (5) Common kestrel (*Falco tinnunculus*) hunting; (6) Natterjack toad (*Bufo calamita*); (7) Field vole (*Microtus agrestis*); (8) Meadow grasshopper (*Chorthippus parallelus*); (9) European rabbit (*Oryctolagus cuniculus*). (D) Shrub phase: (10) Willow warbler (*Phylloscopus trachilus*); (11) Chaffinch (*Fringilla coelebs*).

makes life easier for other plants. The organic matter in the soil also enables it to retain many of these elements so that they are not rapidly lost by the process of leaching. Some of the microbes which grow in the soil may be able to fix nitrogen from the atmosphere, and this adds to general soil fertility.

Animals play an important part in this stage of the development of the succession, since they are often responsible for dispersing the fruits and seeds of later plant invaders. The berries of Sea buckthorn and privet, for example, are eaten by migrating birds, such as members of the thrush family (blackbirds, Song thrushes, redwings and fieldfares) and by some warblers (such as blackcaps). They may carry these considerable distances before dropping them in their feces while sheltering in the vegetation, for instance after a sea crossing. But such birds will not spend time in the developing ecosystem until its vegetation has become attractive to them, as a source of physical shelter, as a means of hiding from predators or as a source of food. So a certain degree of complexity must be attained before animals are a significant factor in successions.

This is true for mammals also. Small mammals, such as voles, shrews and mice, will be found only when the right kind of microhabitat has developed. Rabbits are more mobile, and they invade at a fairly early stage because they find the sandy soil easy to burrow into. They bring with them some fruits which have hooked surfaces and which are dispersed by sticking onto fur, such as the fruits of the Hound's tongue and the burdock. These herbivorous animals have a further effect on succession by being selective in what plants they eat, thus reducing the abundance of the more palatable species.

As the surface of the soil stabilizes, mosses and lichens can establish themselves, and some of these lichens can also fix atmospheric nitrogen. The layer of these lowly plants near the ground surface provides a humid environment in which detritivore animals such as springtails (Collembola) and woodlice (Isopoda) can thrive. This humid carpet is also ideal for the germination of many plant species which, without such protection, could not survive as seedlings in the inhospitable and unstable sand.

When scrub and woodland is established, the resulting more complex vegetation structure gives far more opportunity to animals. Invertebrate grazers have new plant species on which to feed, and so generally occur in much larger numbers. This gives scope for carnivorous animals to make a living, and the numbers of carnivores, both invertebrates such as spiders and vertebrates such as insectivorous birds, increase.

Secondary successions do not show the full sequence of primary successions simply because the starting point is so different. If an area of forest is cleared by man and used for agriculture, and is then abandoned, the starting point of the resulting succession is in no way as extreme as that of the sand dune. The soil is already rich in organic matter and probably contains a good supply of nutrients as well. Many of the bacteria and fungi always associated with a healthy soil will be present already, as will the invertebrate detritivores such as earthworms. This means that the succession can follow a somewhat different path, for it is possible that some of the climax species can invade even at an early stage. Oak trees, for example, may be able to establish themselves within a few growing seasons of the abandonment of a plot of arable land, provided that a source of seeds is accessible.

This does not mean that there will be no sequence of plant and animal species, leading up to the development of stable vegetation. The sequence, however, may be more closely related to the efficiency of invasion, the degree of persistence, the longevity and the generation time of the species involved. For example, abandoned agricultural land will be invaded by weed species within a matter of weeks. Weeds are by nature invasive species. They have well-developed properties of seed production and dispersion, and their seeds often remain dormant for tens or even hundreds of years in the soil. So they always lead the way in a secondary succession. Perennial species, including grasses, broadleaved herbaceous plants, shrubs and trees, may all invade together, but rapid vegetative spread on the part of the herbs may provide them with a short phase of dominance. Dense thickets of shrubs and tree saplings follow, and within them a fiercely competitive battle is fought. Birch trees may come out on top temporarily, but their lifespan is short even though their initial growth is rapid. Eventually they must succumb to the climax dominants, chiefly oak and beech in northwest Europe, spruce and pine in Scandinavia and beech and maple in the northeastern USA.

The actual course of a succession is clearly dependent upon what species of plants and animals are available in the area, but there are certain features which most successions have in common. The most general rule is that they tend to develop towards the ecosystem with the highest biomass possible under the given conditions. But there is an exception even to this: in certain bogs a low-biomass ecosystem dominated by moss takes over from swamp woodland. Another common feature is that the gross energy productivity of the system increases

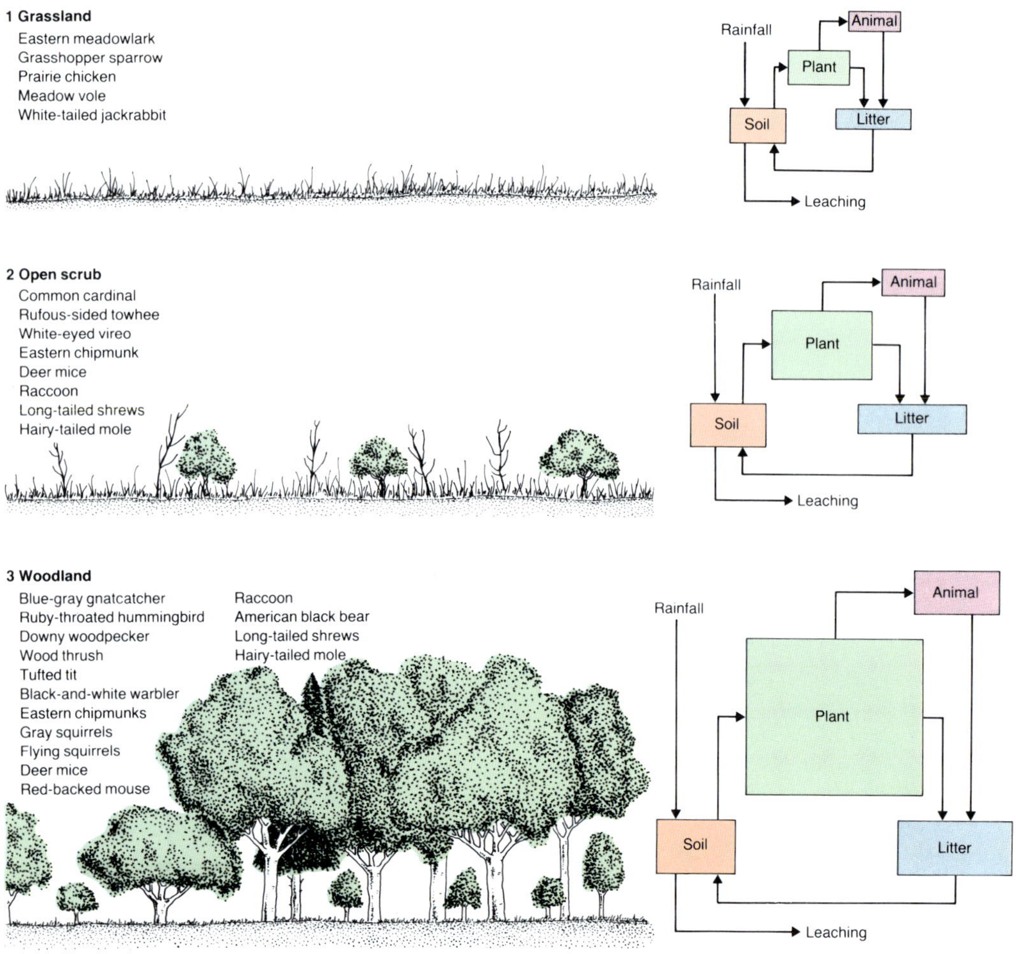

► **Inhabitants of European deciduous woodland.** ABOVE Common dormouse (*Muscardinus avellanarius*), which prefers secondary growth areas where the trees have edible seeds, eg hazel and beech. BELOW Blue tit (*Parus caeruleus*), which often nests in oak woodlands where it mainly feeds its nestlings on the immense caterpillar populations of oak trees.

◄ **Increasing structural complexity and changing diversity** with time—a diagram illustrating the main stages of succession from grassland to woodland in North America and the changing species diversity and pattern of nutrient cycle. The important points to note are that as the ecosystem becomes structurally more complex the fauna and flora becomes more diverse and the reservoir of nutrients moves from the soil (in grassland) to living things (mainly plants) in woodland.

during the succession. But the total amount of energy consumption also rises as the animal and decomposer components of the ecosystem respond to the extra energy input. It is in the early stages of succession that the energy fixed in the ecosystem far exceeds the amount of energy used up, and this surplus energy goes into ecosystem growth in biomass. When man wants to exploit an ecosystem for food production, it is usually found that more energy can be gained in a given time from a young ecosystem, such as an annual cereal crop, than from a mature one such as woodland.

The nutrient cycle of the ecosystem varies with its state of development. At first, the nutrients entering the ecosystem are mainly lost again by leaching, but as the biomass begins to grow, more and more of the nutrients are trapped and held by the living plants and animals. When the climax stage is reached, however, and there is no further net growth in biomass, the total losses of nutrients from the ecosystem will once again equal its total gains, but it will now contain a much greater nutrient reservoir.

The structural and microclimatic complexity of ecosystems increases as they develop, which leads to a greater diversity of animal life. In the case of plants, however, the highest diversity often occurs just prior to the climax state. If the investigation covered all living things, including microbes, then the climax state would probably prove to be the most diverse, but as yet there have been no such complete studies.

There are similar arguments about whether the climax ecosystem is the most stable. Some ecologists define stability in terms of how quickly an ecosystem recovers from disturbance, but this is not very appropriate, since a very simple one such as a field of weeds is more stable in this sense than a complex one such as a forest. A better definition of stability involves the degree to which the ecosystem can resist disturbance, but even here there is considerable disagreement among ecologists concerning what features of the ecosystem render it stable. A complex ecosystem, with a high diversity of plants and animals, usually has highly involved and complicated food webs. But does this make it more stable or more fragile? The general consensus of opinion is that diversity does not necessarily lead to stability. This is an important point for conservationists, since the world's complex ecosystems, such as tropical rain forests, are essentially fragile.

PDM

Regions of the World

HAVING considered the complexity of inter-relationships between organisms and their natural setting, we are now in a position to examine the way in which plants and animals are distributed across the face of the globe. Are there detectable patterns in the assemblages of living things which are found on each of the continents and in the oceans? If so, how are these patterns to be explained?

There are two main ways in which we can approach these questions. The first is to look at the evolutionary history of life and the changing arrangement of the continents over geological time and see whether the patterns of living things can be explained by reference to their origins and history. We can call this the zoogeographical approach. The alternative is to use a strictly ecological approach and see whether the specific ecological requirements of species provide an explanation for their present distribution. In practice, as we shall see, both approaches are necessary for a full picture of the ecological structure of this planet to emerge.

◄ **Asleep on the upper floor**—a Mountain gorilla (*Gorilla gorilla beringei*) resting on a bough in the mountain cloud forest of Zaire.

ZOOGEOGRAPHIC REGIONS

What are they?... The effect of isolation and climate on animal populations... Australasia... South America... North America... Europe and Asia... Africa... The Far East...

EACH of the world's continents has its own unique animals, which are *endemic* to it (that is to say, are known nowhere else). These endemics include the kangaroos and the koala of Australia, the zebras and wildebeest of Africa, and the sloths and armadillos of South America. The study of zoogeography is the attempt to understand the ways in which these faunas came into existence. Each is characteristic of a particular area, called a *zoogeographic region*. Most of these regions are made up of a single continent, such as North America, South America, Africa or Australia. That is because each of these areas is more or less completely isolated from the others by an expanse of sea or ocean. The exception is the Oriental region, made up of Southeast Asia, whose isolation is partly due to the icy Himalayan mountains, which effectively bar it from the rest of Asia. Deserts, too, help to isolate the Oriental region, and also to isolate the fauna of Africa from that of Eurasia.

Just as it is isolation that produces and perpetuates the differences between the faunas of the six zoogeographic regions, so it is mainly the length of isolation that determines just how different each has become. That is why the fauna of Australia, which has been almost completely isolated for most of the last 90 million years, is the most unusual. South America runs second in its length of independence, having been mainly isolated from about 65 million until 5 million year ago, and Africa from about 70 million to about 17 million years ago.

The physical isolation of these continents was due to the process known as *plate tectonics* or *continental drift*. The basic cause of this is the great heat of the earth's core, which creates a pattern of convection currents in the mantle. The earth's crust consists of a number of large, semirigid plates, moving relative to each other as they are transported by the convection forces. Thus over geological time continents move apart (with new ocean floor being formed) or join up (with mountains usually being formed).

The other great influence on the world's faunas has been climate. It is mainly faunal poverty that distinguishes the faunas of the Nearctic and Palearctic regions from those of the warmer lands to the south. Until some 40 million years ago, even the polar regions were free of ice, but the climate then gradually deteriorated, culminating in the series of ice ages of the last few million years. These colder climates are clearly capable of supporting fewer species of animal.

The differences between the mammalian faunas of the six zoogeographic regions will be understood most clearly if the oldest and most distinct are examined first, commencing with that of the **Australian region.**

Australia's isolation started about 90 million years ago, when shallow seas separated it from Antarctica. That continent was still free of ice, and marsupial types of mammal (whose young are born at an early stage of development and complete their development and early growth in the mother's pouch) had used that continent as a route between Australia and South America. Australia itself soon separated from Antarctica, and began its northward journey into the Pacific Ocean. This took place before the continent had been reached by the placental mammals (whose young undergo all their early development within the mother's uterus).

Without the restraint of competition from the placental mammals, the marsupials of Australia were able to diversify and to colonize all the different ecological niches available. They evolved into types very similar, in both appearance and habits, to the dogs, cats, bears, moles, mice and squirrels. Only the kangaroos are very different from their placental counterparts, for most of them are grazers, feeding on the grasses of the endless Australian plains, and therefore taking the place occupied by horses, deer, gazelles and similar herbivorous animals in placental faunas. The Australian mammal fauna also includes the only surviving egg-laying mammals, the Duck-billed platypus and the spiny anteaters (echidnas) which are known nowhere else.

Marsupials are especially well-suited to Australia. That continent has very few mountains, and these are all old and lie along the eastern seaboard, where they receive nearly all of the rainfall from the prevailing easterly winds. Most of Australia therefore consists of level plains, whose soils lack replenishment from the erosion of new mountains and are therefore low in minerals. The resulting sparse vegetation, with tough leaves and poor growth, together with the periodic droughts, provides an inhospitable environment. The ability of the marsupials to slow down or even terminate the growth of the young in the pouch in time of environmental stress is a useful strategy that is not available to their placental relatives.

About 5 million years ago, Australia's northward journey brought it into a chain of islands in the southwestern Pacific. These islands provided a route by which a few of the placentals of Asia were able to colonize Australia. Bats, and later rats, entered that continent naturally, but aboriginal man brought the ancestor of the Dingo dog, while white settlers brought the rabbit and other placentals that now compete with the native Australian marsupials.

Marsupials are also ancient inhabitants of the **Neotropical region**. Over 60 million years ago, South America had been colonized by marsupials, which seem to have evolved somewhere in the Southern Hemisphere. But they had been joined in South America by a few early types of placental mammal, which had evolved in Asia and spread southward through North America. Today, these early types of Neotropical mammal are represented by the marsupial opossums, and by three groups of placentals—the anteaters, armadillos and tree sloths.

The next phase in the faunal history of South America began about 35 million years ago, when the ancestors of the New World monkeys and of the distinctive South American type of rodent appeared in that continent. The monkeys, which have prehensile tails, include capuchins, howlers, spider monkeys and marmosets, while the rodents include Guinea pigs, New World porcupines and the world's biggest rodent, the capybara. It is still uncertain whether these two groups entered South America from North America or from Africa. Whichever was their homeland, they seem to have traveled over a hazardous island route, for no other mammals succeeded in accompanying them.

South America's drift took it progressively westward, until it approached one of the great oceanic trenches where old ocean crust returns into the depths of the earth. This led to the rise of the Andean mountains, whose snowy peaks trap much of the rain from the Pacific winds, so that deserts grew in their lee. But similar volcanic activity also led to the formation of the Panama isthmus, bridging the gap between the Americas. Over this bridge flooded many types of North American placental—foxes, weasels, bears, jaguars, tapirs, llamas, horses and mastodont elephants (although the last two later became extinct in both the Americas). These animals competed

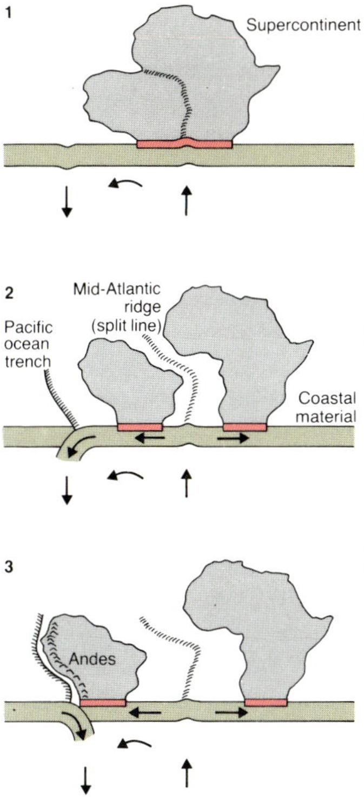

▲ **How Africa and South America drifted apart.** (1) An upward convection current from the deep layers of the earth appeared below the original supercontinent and a downward current in the west. (2) A split appeared above the ascending current and new coastal material, derived from the floor of the new South Atlantic, was formed on either side of the split line, which now formed the mid-Atlantic ridge. An ocean trench formed above the descending current. (3) South America was now adjacent to the ocean trench; movement of coastal material below South America caused earthquakes that created the Andes.

▼ **Worlds apart.** Islands support very individual species, often confined to their island localities (endemic). Shown here is the Barrington land iguana (*Conolophus pallidus*) unique to the Galapagos Islands.

▼ **The world 60 million years ago.** Because of lower sea-levels, and because the North Atlantic had not yet extended northwards, North America was still connected to Europe via Greenland. South America was close to Africa, but, like Australia, had already separated from Antarctica. India had not yet collided with Asia.

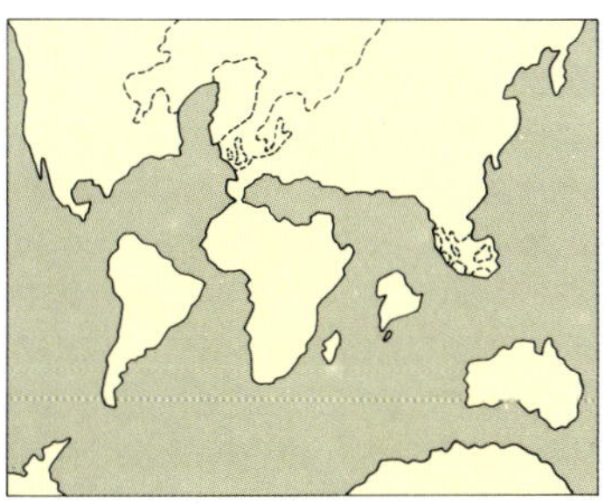

successfully with the native South American mammals, many of which became extinct. As a result, the South American mammal fauna today contains far fewer unusual, unique mammals than before the Panama isthmus was formed. Only three of the South American mammals successfully colonized North America: anteaters, armadillos and opossums.

North America itself, the **Nearctic region**, seems to have been the original home of very many of the families of placental mammal. Most of these were able to disperse eastward, via Greenland and Scandinavia, into Europe. Even after the widening Atlantic closed that route for faunal exchange, lower sea levels left a dry-land connection westward from North America to Asia via Alaska and Siberia. As a result, the faunas of the Nearctic and the Palearctic (Eurasia) were broadly similar for a long time. It was only the increasing climatic severity of the ice ages that gradually closed this route, together with the spread of the shallow Bering Sea. It is significant that the last types to make the crossing between the two regions were hardy mammals such as the elk, moose, caribou, Mountain goat and Mountain sheep (Bighorn). Because of these faunal connections with the Palearctic, and also across the Panama isthmus with the Neotropical region (with which it shares the pig-like peccaries), there are few unique mammal families in the Nearctic—only the pronghorn and some rodents.

The **Palearctic region**, made up of Europe and the main mass of Asia, has had a faunal and climatic history very similar to that of the Nearctic. These two regions are sometimes referred to jointly as the Holarctic, which is the only home for such mammals as pikas and beavers. The Palearctic is separated from the tropical Ethiopian and Oriental regions by the Mediterranean Sea, the deserts of the Middle East and Gobi, and the Himalayan mountains. Were it not for these barriers, there would probably be a much more gradual transition between the more warmth-loving faunas of the Ethiopian and Oriental regions and those of the Palearctic. This was certainly so before the ice ages, when rhinos, hippos, and hyenas lived in Europe. The fauna of the Palearctic today, impoverished by the ice age extinctions, has only a few types of unique rodent, and the unique Giant panda.

Africa, the **Ethiopian region**, was separated from Eurasia by shallow seas until about 17 million years ago. A few types of early mammal nevertheless reached Africa, so that an "Ethiopian" fauna evolved there. This included the elephants and the related hyraxes, as well as the little elephant-shrews and golden moles. Early ancestors of the Old World monkeys and the apes also seem to have diversified in Africa. When the barrier seas withdrew, the Ethiopian fauna was able to spread into southern Eurasia, and there is still a great deal of similarity between the faunas of the Ethiopian and Oriental regions. Because they are no longer found in the now-cooler Palearctic region, such mammals as the Old World monkeys, apes, porcupines, rhinos, elephants and pangolins are now found exclusively in those two regions, as are those small relatives of the monkeys, the lemurs and lorises. Nevertheless, the faunas of the two regions are not identical, for they have become isolated from one another by the deserts that resulted from the increasing dryness of the Middle East, and by the opening of the Red Sea. As a result, elephants, rhinos, apes and porcupines have evolved into slightly different types in the two regions—for example, the African elephant, White rhino and chimpanzee or gorilla differ from the Indian elephant, Javan rhino and orangutan.

The fauna of the Ethiopian region also became more distinct because East Africa became drier, so that grasslands expanded there. Many types of grazing animal evolved to feed on these grasslands—antelopes, gazelles, gnus, impalas, zebras, giraffes, buffalos, warthogs and so on.

The fauna of the **Oriental region** is, then, very similar to that of the Ethiopian region. Today, its tropical fauna seeps northward into China around the eastern edge of the Himalayas and, as noted earlier, is gradually extending eastward across the East Indian islands toward Australia. Its unique mammals are the

gibbons, tarsiers, tree shrews and flying lemurs or colugos (which really glide rather than fly, and are not related to lemurs!). Most of the Oriental region is made up of the Indian subcontinent, which had an unusual history. It began as part of a great southern landmass, adjacent to Africa, Antarctica and Australia. After splitting away from them over 130 million years ago, it drifted northward to collide with Asia about 45 million years ago; the consequence of this collision was the raising of the Himalayan mountains that now separate its fauna from that of Eurasia. It would be fascinating to know the details of the changing history of the Indian fauna, but unfortunately that part of the Indian fossil record still remains to be discovered. CBC

▲ **Zoogeographic regions** of the world and some of the characteristic or shared animals. Palearctic/Nearctic: (1) Red deer or wapiti (*Cervus elaphus*). (2) Red fox (*Vulpes vulpes*). Nearctic: (3) Pronghorn (*Antilocapra americana*). (4) Mountain beaver (*Aplodontia rufa*). Neotropical: (5) Woolly opossum (*Caluromys lanatus*). (6) Jaguar (*Panthera onca*). (7) Capybara (*Hydrochoerus hydrochaeris*). (8) Brown-throated three-toed sloth (*Bradypus variegatus*). (9) Brown capuchin (*Cebus apella*). African or Ethiopian: (10) Gorilla (*Gorilla gorilla*). (11) Giraffe (*Giraffa camelopardalis*). (12) Ringtailed lemur (*Lemur catta*). Australian: (13) Tasmanian devil (*Sarcophilus harrisii*). (14) Eastern gray kangaroo (*Macropus giganteus*). (15) Brown antechinus (*Antechinus stuartii*). Oriental: (16) Orang-utan (*Pongo pygmaeus*). (17) Terrestrial tree-shrew (*Lynogale tana*). (18) Lar gibbon (*Hylobates lar*). Palearctic: (19) Giant panda (*Ailuropoda melanoleuca*).

THE WORLD'S BIOMES

The effect of latitude on climate. . . The movement of air masses. . . The influence of land and sea. . . Similar climates produce similar life-forms. . . Examples of biomes, grasslands. . .

It is very evident that the different kinds of plants and animals found on earth are not distributed in a random way over its surface. This is because of the pattern of climates over the globe, which determines what kinds of animal and plant are able to survive. There are many aspects of the earth's geography which affect the climate, for instance climatic conditions vary with latitude. Equatorial (low) latitudes are generally hotter than polar (high) latitudes, which is a consequence of the angle at which radiant energy from the sun arrives at the earth's surface. Energy arriving at the equator is spread over a smaller surface area than the same amount of energy arriving at the poles; it also passes through less of the atmosphere, so the degree of energy absorption within the atmosphere is lower.

The overall difference in temperature at the equator and at the poles also has an influence upon the global circulation of air masses. Cooler air at the poles is dense, so it sinks and spreads toward the equator along the earth's surface; converging masses of cool air at the equator force upward the warm, low-density air mass which develops around the hot equatorial regions. The result is a global pattern of air movements in which air rises over the equator and sinks over the poles; but the system is made more complex by a reverse circulation pattern over the mid-latitudes. The light, warm air mass over the equator is cooled as it is forced upward and it descends again around 30° north and south. From here some moves back to the low latitudes and some moves poleward, where it meets the polar air mass coming in the opposite direction, resulting in an unstable zone of colliding air masses.

These air-mass movements have a considerable effect on global climate patterns. Equatorial regions with rising air have low-pressure conditions and the hot, moist air, cooling as it rises, sheds much rain. Around 30° north and south are high pressure belts, created by descending air; these are generally arid regions. Around 50–60° north and south is the unstable region, bringing alternating high- and low-pressure systems, often with plentiful moisture. The polar areas have high pressure with descending air masses, resulting in very low precipitation, hence the "polar deserts."

The pattern of surface winds over the earth also influences climates, since these may carry moist air when they come off the sea and dry air when moving over extensive land masses. The winds are not generally directly toward the poles or the equator, but are deflected by the rotation of the earth, to the right in the northern hemisphere and to the left in the southern hemisphere (the *Coriolis effect*).

The pattern of climates which emerges from these processes is further complicated by the arrangement of land and sea. Land masses heat up faster and cool down faster than the sea, so that there are "continental climates" with high summer and low winter temperatures in the interior of the land masses and "oceanic climates" with a low amplitude of variation around the coast. The coastal areas may further be affected by cold

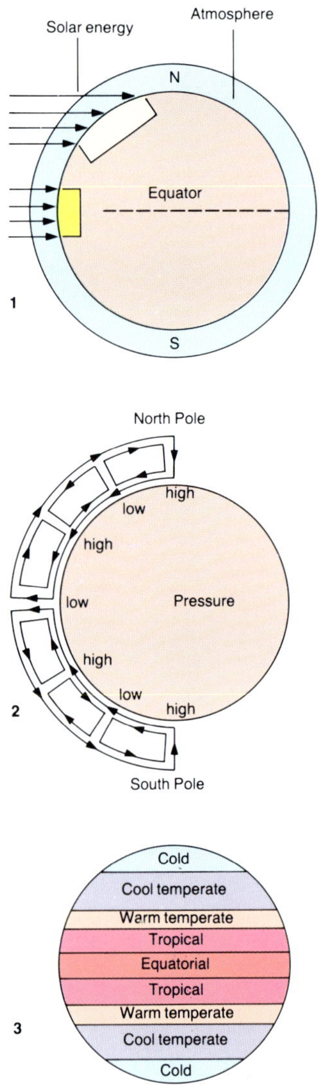

► Environmental factors at work. The life-forms present in any habitat are the result of the interactions of the physical and natural world. Various factors are at work in this savanna landscape in the Serengeti Plains, Tanzania. Browsing by ungulates and a prolonged dry season with fires restrict the tree cover. Here a thunderstorm is about to end the dry season, bringing the rains that sustain the savanna grassland.

◄ Factors affecting the earth's climate. (1) Arrival of solar energy at the earth's surface. Note that the energy arriving at the equator is spread over a smaller surface area than is the same amount which arrives at the poles. (2) General circulation pattern of air masses in the earth's atmosphere. Note that an overall pattern occurs in which air rises over the equator and sinks over the poles, upon which is superimposed a reverse circulation pattern over the mid-latitudes. (3) Summary of the world climate zones.

▼ Plant life-forms in communities. A summary of five vegetation types showing how the plant types (life-forms) vary with location and climate.

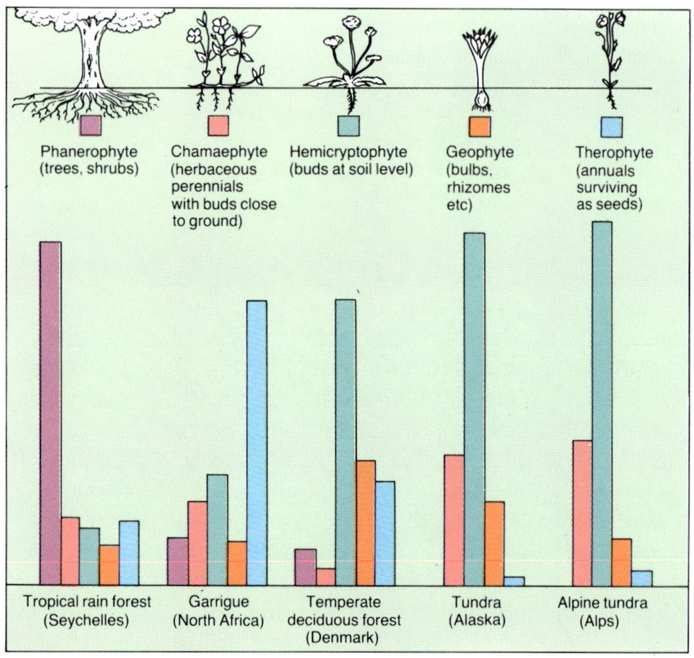

or warm ocean currents impinging upon them. Altitude is also of great importance: high elevation often means lower overall temperature and higher precipitation, and high mountains can also affect surrounding regions by shielding them from moist air movements, as the Himalayas shield the Gobi Desert from monsoon rains.

The complex mosaic of climates across the face of the earth has led to an equally complex development of vegetation types (or *formations*) and these in turn support different kinds of animals. These major, global units of flora and fauna are termed *biomes*. Their distribution pattern relates closely to the climatic patterns produced by all of the variables which have just been discussed, and the organisms present show adaptations which enable them to survive under the climatic conditions prevailing (temperature and water availability being the most influential factors). The exceptions are the marine habitats in which depth (and therefore pressure) together with ocean currents are of prime importance, and the man-made habitats (the agricultural and urban habitats) where the effects of man assume a preeminent role in determining flora and fauna.

For the natural, undisturbed terrestrial biomes, one can discern a similarity in the general appearance and structure of the vegetation and often of the animal life, despite geographical separation and taxonomic variation. For example, the tropical grasslands of South America are very similar in form to those of central Africa, although very different species are present. So biomes are characterized by the "life-form" of their vegetation rather than the precise species found there.

The Danish botanist Christen Raunkiaer gave a precise formulation to the concept of life-form. He defined a series of life-forms in plants based upon the height above ground at which their *perennating organs* (buds, bulbs, corms, tubers, etc) were held. Under warm, moist conditions, there is no need to protect these structures, hence they can be held high in the air. Thus in the tropical, low-pressure zone (bearing the tropical rain forest biome) many types of tree are found. Under drought conditions, plants have their buds closer to the ground, or may even survive underground (*geophytes*) during unfavorable periods, or, alternatively, may survive as seeds (*therophytes*). In the tundra biome, the air is cold and winds may bear ice crystals, yet below ground the deeper soil layers are permanently frozen. Only in a zone just above and just below the soil surface can plant life survive: such dwarf cushion plants as survive here are termed *chamaephytes*.

Biomes, therefore, have a characteristic spectrum of plant life-forms. They also have animals which occupy particular roles within the ecosystem, tapping particular environmental resources. These may differ very considerably in their taxonomy from one part of the world to another, but are nevertheless "ecological equivalents." For example, the South American pampas is grazed by the guanaco, which is equivalent to the Australian kangaroo, the Asiatic ass and the North American bison in that it is a relatively large, fast-moving herbivorous animal living in herds. There are many such examples of an animal being replaced by a very different one, but one which occupies the same functional niche in the equivalent biome in another part of the world. PDM

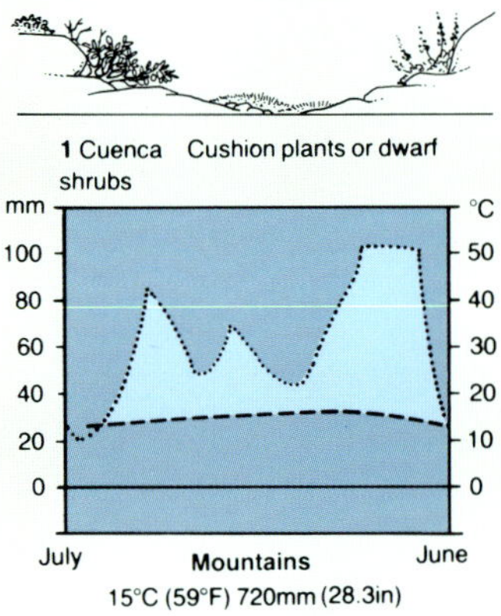

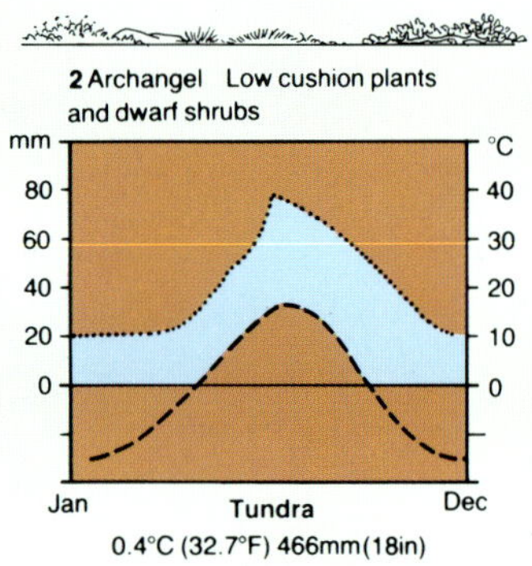

▲► **The world's ten major biomes.** A summary box is provided for each biome. These provide a background-color key to the map; the line sketch indicates typical vegetation and the graphs annual rainfall and temperature (see key) for particular locations indicated by numbers on the graphs and map.

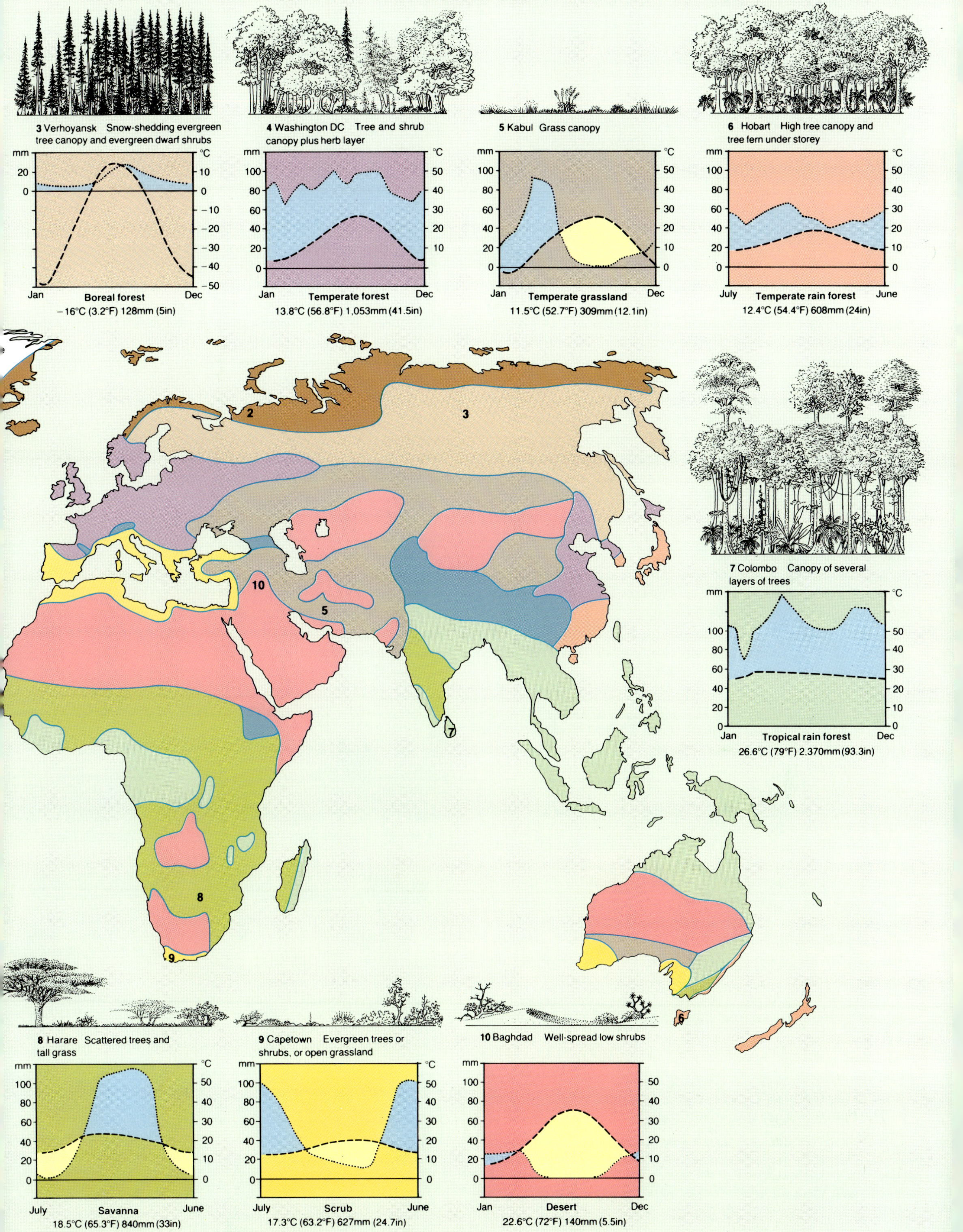
3 Verhoyansk Snow-shedding evergreen tree canopy and evergreen dwarf shrubs
Jan Boreal forest Dec
−16°C (3.2°F) 128mm (5in)
4 Washington DC Tree and shrub canopy plus herb layer
Jan Temperate forest Dec
13.8°C (56.8°F) 1,053mm (41.5in)
5 Kabul Grass canopy
Jan Temperate grassland Dec
11.5°C (52.7°F) 309mm (12.1in)
6 Hobart High tree canopy and tree fern under storey
July Temperate rain forest June
12.4°C (54.4°F) 608mm (24in)
7 Colombo Canopy of several layers of trees
Jan Tropical rain forest Dec
26.6°C (79°F) 2,370mm (93.3in)
8 Harare Scattered trees and tall grass
July Savanna June
18.5°C (65.3°F) 840mm (33in)
9 Capetown Evergreen trees or shrubs, or open grassland
July Scrub June
17.3°C (63.2°F) 627mm (24.7in)
10 Baghdad Well-spread low shrubs
Jan Desert Dec
22.6°C (72°F) 140mm (5.5in)
mm
°C
2
3
5
6
7
8
9
10

TROPICAL RAIN FOREST

Climatic conditions. . . The most productive biome. . . Typical vegetation. . . Soil types. . . Animals above the canopy, within the canopy, below the canopy and on the forest floor. . . Army ants. . .

"TROPICAL rain forest" conjures up images of dense, impenetrable jungle with lush undergrowth. In fact this is not a true picture of virgin forest, for there one's passage is not usually impeded by a great density of growth. Perhaps the erroneous impression has arisen from travelers' tales, based either upon disturbed forest which is in the process of recovery, or on riversides where there is a great density of low-growing shrubs. Much depends on the penetration of light. If the light penetration is good then there will be much undergrowth, but this only occurs in old clearings and riverbanks.

The tropical rain forest, as its name implies, is found largely in equatorial regions and is almost entirely confined to the tropics. Here it receives high rainfall because of the convectional nature of the air currents: warm, humid air masses rise and are replaced by the denser air of the trade winds converging. The total rainfall usually lies between 2.5 and 4m (100–160in), and there are sites where 10m (400in) of rainfall have been recorded in a single year. In the true rain forests this is distributed fairly evenly over the whole year, but as one moves away from the equator there is an increasing tendency for the forest to experience dry and wet seasons. This affects the type of vegetation which can develop.

Temperatures are also generally high and show little seasonal variation. Usually the temperature of the lowland forests lies between 20° and 28°C (68°–82°F).

Equatorial regions have not always enjoyed a tropical climate. When the high latitudes were experiencing the ice age that reached its greatest extent some 18,000 years ago, the equatorial regions were drier than they are today and the rain forest was probably much reduced in area, perhaps being fragmented into relatively small relict areas.

Despite the frequent cloud cover over the forests, light intensities are consistently high throughout the year, and this generates a constant high temperature. The combination of warmth and wetness makes the habitat extremely productive. The net primary productivity often exceeds 2kg/sqm/year (0.5lb/sqft/year) and can reach values of 3.5kg/sqm/year (0.9lb/sqft/year). This places the rain forest among the most productive biomes in the world; its productivity can only be matched by some agricultural systems and some swamps, estuaries and algal ecosystems.

Unlike its competitors in productivity, the tropical rain forest has an extremely high level of biomass, often 45kg/sqm (11lb/sqft) and sometimes as high as 80kg/sqm (20lb/sqft). It is this vast bulk of timber, together with the demand for agricultural land, which has led to the extensive exploitation of these forests around the world.

The biomass is organized in a complex system of layers, the precise arrangement of which is highly variable. In undisturbed forest two or three layers of tree canopy can often be discerned. The highest or *emergent* trees are often 35–40m tall (115–130ft) and form an open, broken canopy above the rest of the forest. The next layer is usually much more continuous and lies at about 20m (65ft), and below this, around 15m (50ft), there is often a third layer. The lower and middle layers are not always clearly separable.

Within such a forest very little light penetrates to the floor, and hence the growth of underscrub and saplings is seldom very dense. The atmosphere is permanently dark, humid and still. Many herbaceous plants manage to find a roothold in the upper branches of trees and exist there as *epiphytes*, dependent on other species for support. Creepers are abundant, straggling up to the light. Some trees, such as some types of fig, overcome this problem of having to establish themselves in the darkness of the forest floor by germinating in the crevices of the bark of other trees and existing as epiphytes in their early years. They then send down roots to the soil and may eventually obliterate their host as they outgrow it.

In the soil, decomposition is rapid. So much litter falls to the ground, supplying energy to the detritivores and decomposers, that many herbaceous flowering plants have evolved into *saprophytes* (living on the debris) and have lost their green pigment. Energy is easier to obtain secondhand than it is by photosynthesis in this dark habitat.

The soils are generally poor in nutrients because of the high rainfall and rapid washing out (leaching). Nutrients released from litter decomposition are rapidly taken up by the growing trees. This means that cleared rain forest does not produce good agricultural land, although it may be productive for a few years following its initial burning. Soil fertility and stability is soon lost under the heavy rainfall, and agricultural projects in many areas of cleared forest throughout the world are having to be abandoned as too costly.

The tropical rain forest is the home of some of the most specialized of animals and has the greatest species richness of any

▲ **Drinking party.** A group of butterflies, *Graphium sarpedon*, sucking up water and its dissolved nutrients from the moist sandy soil in a clearing in Malaysian rain forest. The insect life of tropical rain forests is spectacular, and many thousands (even millions) of species probably remain to be identified.

► **"Arboreal cow"**—a Brown-throated three-toed sloth (*Bradypus variegatus*) searches for its leafy diet in an emergent tree in the Panamanian jungle. Sloths have many-compartmented stomachs which contain cellulose-digesting bacteria, just like those of more conventional herbivores such as cattle.

biome. Among the emergent trees, there is a community of animals that rarely if ever venture down into the lower levels. Their world is bright, windy and warm. Some of the largest birds of prey live here. The Harpy eagle is the world's largest eagle, soaring over the forests of South America to prey on monkeys and sloths. In the African rain forests the same ecological niche is filled by the smaller Crowned eagle. Further eastward, the Philippine Monkey-eating eagle is the top carnivore in the emergent layers of the rain forests of Southeast Asia. These birds need large tracts of forest to provide their food. They are slow-growing and produce only one young at a time. These large birds of prey are some of the first to suffer when the rain forest is disturbed, since removal of the large trees takes away the birds' habitat, and fragmentation of the forest breaks up the large tracts over which they need to forage.

The emergent region is also occupied by the bizarre but beautifully marked toucans and hornbills. The toucans are found in the neotropics and are fruit-eaters. Their large, brilliantly colored beaks are thought to be identification flags. They are weak flyers, and move about in small groups. The hornbills of Africa and Southeast Asia have a similar lifestyle.

Descending into the shade of the rain forest canopy, the variety of animal life abruptly increases. Sound becomes an important method of communication as vision is impeded by the dense vegetation. The canopy forms a highway for monkeys and sloths, while smaller animals have developed some remarkable methods to glide from branch to branch. The Paradise tree snake flattens its ribs to turn its normally cylindrical body into a curved ribbon which helps it to glide through the air. Wallace's flying frog has enlarged webbed feet which are used in gliding. Some of the smaller mammals, such as the flying lemurs of Southeast Asia, also glide, using flaps of skin along their sides. These flaps are stretched out by spreadeagling their legs. Larger and more common are the monkeys and primates, which have exceptionally well developed hands and feet to enable them to swing between the canopy branches.

Birds are very obvious in and around the canopy. Parrots are found in all the main rain-forest areas, moving about in noisy, brightly colored groups. Man has a great fondness for parrots as pets, and despite legal protection they are still being captured, exported and sold at vast prices. In the rain forests of Papua New Guinea the birds of paradise make the brilliance of the parrots look positively dull. With stunningly iridescent colors and long tail streamers the paradise birds cavort and display to their mates. The Blue bird of paradise hangs upside down, puffing out its shimmering chest feathers and calling in a deep reverberating voice.

The insect world is no less spectacular. Butterfly larvae eat plants, and the wealth of plant life provides for a great variety of butterfly species. Some of the most beautiful are the birdwing

◄ **Acrobat of the Malaysian jungle**—a Lar gibbon (*Hylobates lar*). Primates have exceptionally well developed hands and feet to enable them to swing between the canopy branches. The most highly specialized of these "swingers" are the gibbons of Southeast Asian rain forests. Their arms are very long (they touch the ground when the gibbon stands upright) and are used in a "hand over hand" action as the animal moves rapidly among the branches. The monkeys of South America have a tail which they can use to hold onto branches. African monkeys' tails do not have this grasping facility. Monkeys are frequently social, moving in small groups, and they often have territories the sizes of which are related to the type of food they eat. Leaf feeders have fairly small territories since their food is plentiful, but fruit eaters have to have much larger territories to ensure that they always contain some trees that are fruiting.

► **Displaying his handicraft,** a brilliant male Regent bowerbird (*Sericulus chrysocephalus*) parades in front of his bower, with the rather drab female behind.

butterflies of Southeast Asia, although their numbers are being reduced by collectors. Many other insects are present but are cunningly camouflaged. Some look like twigs, others like flowers, seeds, leaves or thorns. In a recent study of rain-forest canopy insects, over 600 new beetle species were discovered on a single tree species. If a similar variety of new insect species is associated with each of the very many species of rain-forest trees, then there may be as many as 30 million insect species on the earth, although only about one million are described.

The understory is the domain of the big cats of the rain forest: the jaguar and the ocelot in the South American jungles, the Golden cat in West Africa and the Clouded leopard in Southeast Asia. They take a variety of small prey including birds, rodents and reptiles. The larger species wait silently in the branches and drop onto ground-dwelling tapirs or peccaries.

Within the rain forest the high and constant humidity provides conditions suitable for species more commonly associated with aquatic habitats. Tree frogs are common and are often very poisonous, a fact which they advertise by being brightly colored. They lay their eggs in water that collects in tree holes or in epiphytic bromeliads. Another group of animals are the land planarians—flatworms which move along slime trails by the beating of millions of small hair-like cilia. Land planarians are brightly striped carnivores and hunt at night; they may reach a length of almost 600mm (2ft).

The largest occupants of the rain forest live down on the forest floor. Elephants, tapirs, okapi and gorillas all feed on the understory vegetation while the very sparse ground cover provides food for chevrotains (small deer) in Africa and Southeast Asia, and for their rodent equivalents, the pacas and agoutis, in South America. These herbivores are prey to the big cats and also to other forest-floor dwellers. The fer-de-lance and the bushmaster, two of the world's more poisonous snakes, feed here on rodents and birds. The forest floor supports a great variety of birds including the Jungle fowl, from which the domestic chicken is derived, as well as several pheasant species. In South America the ecological equivalents of the fowls and pheasants are the curassows and tinamous.

The forest floor is also used as a display ground for the bowerbirds and the cocks-of-the-rock. Male bowerbirds build elaborate constructions, gaily adorned with berries, flowers and feathers, to attract a mate, while cocks-of-the-rock clear dance floors for their colorful displays. The small neotropical antbirds have a very one-sided relationship with army ants. Army ants have a nomadic way of life (see pp46–47).

The rich jungle fauna and flora are intricately interwoven into a complex but highly fragile ecosystem. Just *how* fragile is only now becoming appreciated. Recognizing the dangers of destroying rain forests through logging is one thing, but it is another matter to prevent it. PDM/BDT

Military Raids on the Forest Floor

The lifestyle of army ants

Army ants are renowned in both the Old and New World tropics for the turmoil caused by their huge raids, the immense size of their colonies and the military precision of their nomadic behavior. Remarkably, there is now evidence that army ants with this set of characteristics have evolved independently in Africa and in the Americas. This convergent evolution can be attributed to the fact that all army ants are predators either of social insects or of large arthropods. Because these prey can only be captured by large numbers of workers foraging in well coordinated groups, army ants have large colonies. For example, colonies of the African driver ant (*Dorylus* species) may have up to 20 million workers. As a result, all army ants exhibit elaborate nomadic behavior, because their massive raids locally deplete their prey populations, which are slow to recover, and hence the entire colony must frequently move to new feeding areas.

Of all army ants the ecology and behavior of *Eciton burchelli* colonies is best known. This species inhabits lowland tropical rain forest from Peru to Mexico, and their colonies are remarkable because they have fixed 35-day cycles of raiding and migration that are associated with the growth of new broods of army ant workers. Just as in other ants, new workers develop from eggs that hatch into growing larvae which later pass through a non-feeding pupal stage before they become adults. However, army ants are unusual because their broods of new workers are produced in discrete batches whose development is coordinated with the raiding and nomadic behavior of the colony. The production of discrete broods of workers is possible in army ants because each colony has an immense queen who can lay a large number of eggs in a very short time. In *Eciton burchelli* the queen lays 60,000 to 100,000 eggs at a set time within her colony's activity cycle. These eggs later hatch in synchrony to provide the colony with a brood of larvae which all develop over the same 15-day period. When it has larvae to feed, the colony raids every day and emigrates with its larvae to a new nest site almost every night. This nomadic behavior ends when the larvae spin their pupal cases. Pupal development takes 20 days, and throughout this period the colony remains in the same nest site and raids on only 13 days on average. During this low activity (*statary*) phase in its cycle the colony can raid less frequently because its brood does not require feeding. Halfway through this 20-day statary phase the queen lays a new batch of eggs which hatch at exactly the same time as the new workers are emerging from their pupal cases. Hence at the end of the statary phase the army ant colony has a new platoon of workers to march and a new brood of larvae to feed, and so it enters another 15-day nomadic phase.

In this way a colony increases its population of workers every 35 days. When a colony has about 600,000 workers it still maintains a 35-day activity cycle but may now produce a brood of a few thousand males and about six queens. The males fly away to fertilize queens of other colonies, and the parental colony splits into two new ones. The new queens compete with each other and the old queen for possession of the new colonies. Such young colonies may take three or more years to grow to a size at which they can divide. It is known that individual queens may live more than five years and some workers may live for 300 days or more.

▲ **Raiding party**—a group of the army ants *Eciton burchelli* carrying a centipede back to their bivouac. The smaller, partly black, ants are workers, the larger paler ones soldiers.

▼ **Cycles of behavior**—the 35-day foraging and migration pattern of the army ant *Eciton burchelli*. The spoke lines during the low-activity "statary" phase indicate the direction of raiding parties and the numbers the sequence of raids. Note the angle between each raid is 126°. During the nomadic phase there is a new nest site almost every night.

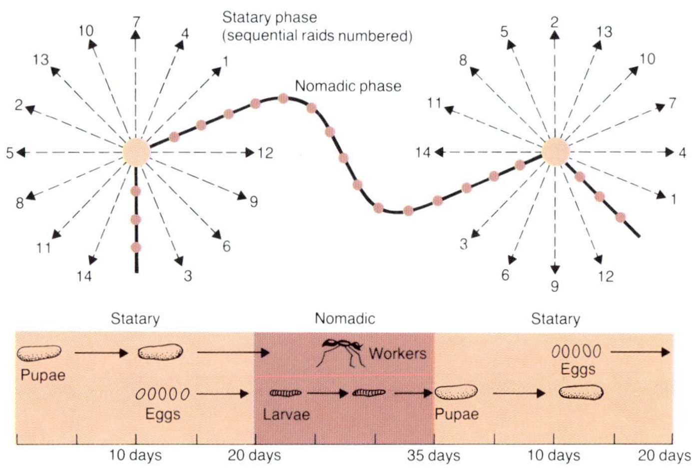

These army ants have remarkable foraging patterns that serve to lower the frequency at which they will encounter areas already raided by their own colony and others of their species. In the statary phase successive raids are directed in a way that gives the ants' prey maximum time to recover their numbers on the forest floor before the next raid. This is achieved by a spiral arrangement of the raids; successive raids are separated by approximately 126°. This is the very same angle at which some plants with fifteen leaves (corresponding to the 13 statary raids plus the nomadic raids leading to and from the statary bivouac) place successive leaves to achieve the minimum of overshadowing. In addition the army ants navigate in the nomadic phase in a way which will help both to prevent the nomadic raid path crossing itself and also to separate the areas foraged in successive statary phases. To achieve these patterns army ant colonies lay down chemical trails, and there is some evidence that *Eciton burchelli* colonies may recognize and avoid the scent-marked trails of other colonies of their species, both to avoid raiding recently foraged areas and to avoid a collision with another colony. This scent marking of foraging trails is functionally similar to the territorial behavior of many species of wild cat.

Eciton burchelli colonies can move easily to new parts of the rain forest because they make bivouac nests that take the form of hanging baskets of living workers. These bivouacs are most often found suspended from the sides of trees; they cradle both the colony's brood and its queen. The army ants in the bivouac regulate the temperature within their living nest by altering how tightly they cluster. Through this behavior the temperature within the bivouac is maintained fairly steadily above external temperatures to provide a suitable environment for the rapid growth of the brood.

The raids of *Eciton burchelli* are the largest of any neotropical army ant; they contain up to 200,000 ants and can be 20m (66ft) wide. The raid moves constantly forwards as a phalanx of ferocious workers, and such is the density of ants in the swarm that they make the floor of the rain forest look like a seething river of marauding insects. The swarm front proceeds at an average rate of 14m per hour (46ft/h) which is sufficiently slow that any terrestrial vertebrate can outrun these predators. However, the raiding army ants capture a wide variety of large arthropods such as cockroaches, spiders and scorpions—though their major prey are other ant species that nest on the forest floor. Larger items are ripped to pieces at the swarm front and carried back to the bivouac.

Raids of *Eciton burchelli* attract a large number of different species of birds, who do not prey on the army ants, but pick up the cockroaches and other large insects that the army ants flush from the leaf litter. A number of these bird species, including the Bicolored ant bird, rely entirely on the army ants for their food supply. Flying insects also attend the army ant swarms: these include parasitic flies that lay their eggs on the insects fleeing the army ants, and certain species of butterfly that follow ant swarms to find the droppings of the ever-present ant birds, from which the butterflies obtain essential nutrients. *Eciton burchelli* colonies also have vast numbers of secretive camp followers. There are *myrmecophilous*, literally "ant-loving," arthropods that live within the nests of the army ants. Thus *Eciton burchelli* colonies provide feeding opportunities for a wide range of tropical species that are entirely dependent on these army ants for their survival.

In addition to these species which follow the army ants, other members of the leaf-litter fauna also gain from the presence of army-ant raids. Recent work on Barro Colorado Island in the Republic of Panama has shown that the raids of *Eciton burchelli* maintain the diversity of the ant species which nest on the forest floor. In the absence of *Eciton burchelli* raids some ground-nesting ants become so abundant that they displace other ant species. However, the army-ant raids remove these competitively dominant species, providing space in which other ants flourish. Indeed, some forest-floor ants seem to specialize in the use of the gaps created in the leaf-litter fauna by army-ant raids; and such gaps are perpetually being opened up by swarm raids.

There are about 50 *Eciton burchelli* colonies in the 1,500 hectares (3,700 acres) of forest on Barro Colorado Island and the raids of these army ants sweep over an area of approximately 1,000 hectares (2,500 acres) each year. It takes a number of months for the social insect fauna to recover from a raid, so *Eciton burchelli* maintains the leaf-litter fauna in a patchwork quilt of recently raided, recovering, and fully recovered areas, and this patchiness helps to maintain the diversity of the rain-forest insects. NF

TEMPERATE RAIN FOREST

Typical vegetation. . . Climatic conditions. . . Animal species in North America, South America, Asia, New Zealand and Tasmania. . .

As one moves poleward from the Equator, the climate undergoes a change and drought at particular seasons becomes more and more apparent. But just outside the Tropics of Cancer and Capricorn, particularly on the eastern sides of the great continental masses and on the islands, where oceanic influences are strong and seasonal factors are dampened, is found the region of warm temperate rain forest. The life-form of the major trees here is the broad-leaved evergreen. In eastern North America there are coastal regions with evergreen oaks, although much of this climatic zone is taken up with swamp forest with the swamp cypress dominant, or sandy areas with pines. In eastern Asia, both on the mainland, such as Korea, and on the islands, such as Japan, evergreen trees of the oak and beech family predominate together with some conifers including the Japanese red cedar.

In the Southern Hemisphere, evergreen forest runs right from Brazil down as far south as the River Plate, and also occupies much of New Zealand, southeast Australia and Tasmania. There is also a west-coast, oceanic version of the warm temperate forests in North America in the form of the great redwood forests which stretch from northern California to the Canadian borders.

Southern beeches are the most characteristic trees of the Southern Hemisphere forests, and in the North Island of New Zealand the coniferous kauri pines occur, together with palms, which provide a link with the tropical rain forests of Malaysia and Sumatra.

The characteristic climate of these regions is wet and mild (but not hot). There is no frost and no dry period, although the bulk of the rainfall, which may be as much as 1,000–2,000mm (40–80in), usually falls in winter. The biomass of the vegetation is high, and some of the tallest trees in the world occur within this biome, such as the Californian redwood, with heights of over 100m (330ft) recorded. The complexity of layering, however, is not as well developed as that found in the tropical rain forest, and the productivity is somewhat less, being about 1.0–1.5kg/sqm/year (0.25–0.37lb/sqft/year). But this is still a very productive biome which is exploited extensively for forestry wherever it is found in the world.

The debris produced by these temperate rain forests, however, in the form of leaf litter, dead twigs and branches, is much more apparent than in tropical rain forests. It lies thick on the ground, not because springtails, woodlice, earthworms and other primary decomposers are absent, but because the cooler climate inhibits the metabolic activities of these cold-blooded animals. For the same reason lizards, snakes and frogs are far less common than in tropical rain forest. The slow rate of decay gives rise to a very congested forest floor, littered with fallen tree trunks and grown over by vines and other vegetation, so that progress through the forest is difficult for large animals. Even so, in the temperate rain forests of North America, such as that in the Olympic National Park, Washington, there are a number of large animals like the Black-tailed deer and the wapiti which browse on vegetation, and the American black bears, which are adept climbers of trees. Other mammals of this National Park include the smallest species of mole, the American shrew-mole, and the American marten.

Forests of other countries have similar types of animals, for example they also have black bears, the Spectacled bear in South America and the Asian black bear in China and Japan.

But it is in the temperate rain forests of New Zealand and Tasmania that some of the most unusual species live. On the forest edges in Tasmania lives the small, shy and secretive Red-bellied pademelon, a common species of wallaby which is hunted for its fine pelt. Within the rain forest there are several other marsupials, including the mouse-sized Eastern pygmy possum and the uncommon Tasmanian Tiger quoll. This quoll is a marsupial that occupies a similar feeding niche to the small cats, climbing trees and catching roosting birds. Several rodents coexist on the forest floor, including the endemic Long-tailed rat and a subspecies of the widespread Eastern swamp rat.

The rain forests of New Zealand are the principal habitat of the kiwi. There are three kiwi species; all are found on the South Island, but only the Brown kiwi is now found on the North Island. They have poor eyesight and find their food by smell; the nostrils open at the tip of the long bill. They smell their food while rooting about in the leaf litter. Kiwis are omnivorous, feeding on insects, seeds and berries.

These New Zealand rain forests, like so many other unique forest types, are under considerable attack, not only from logging but also from species introduced from elsewhere. Somewhat curiously, the rain forests' main menace is not from pigs, goats or sheep but from the Common brushtail possum, which was introduced for its fur. These animals eat the growing tips of many of the forest trees, causing their death. Seedlings are unable to establish themselves because of the grazing pressure of introduced deer and of the opossum. PDM/BDT

◄▼ **Animals of the Olympic National Park,** Washington State, USA. (1) American black bear (*Ursus americanus*). (2) Douglas squirrel (*Tamiasciurus douglasii*) (3) North American porcupine (*Erethizon dorsatum*). (4) Fisher (*Martes pennanti*). (5) Black-tailed or Mule deer (*Odocoileus hemionus*). (6) American shrew-mole (*Neurotrichus gibbsi*).

TEMPERATE DECIDUOUS FOREST

Climate and vegetation... Structure and productivity... Location and extent... Animal populations... Hibernation... Bird migration... Winter survival—escape or hibernation... The ecological cycle—leaf fall and insect life... Spring and summer—bird and insect foodwebs...

JUST as a prolonged dry season in the tropical region results in deciduous trees being at an advantage over the evergreen broad-leaved ones, so in the temperate zone a prolonged cold spell in the winter results in this deciduous type of forest developing. The broad evergreen leaf is very effective so long as it does not need to cope with dryness or freezing, but becomes very inefficient when it is faced with these problems. It may have a subtle advantage over the deciduous leaf if the soils are poor in nutrients, for it may live for several years and act as a storehouse for the scarce nutrient resources. It is also able to commence operating as soon as conditions become favorable. But over much of the temperate area, where winter frost is a regular experience, the deciduous habit has the edge over other survival strategies.

The climate of areas where this biome has developed is characterized by adequate rainfall of about 300–1,200mm (12–48in), fairly evenly dispersed over the entire year. There is no dry season. The summer typically lasts for four to six months and is a highly productive season, while the winter is sufficiently cold for most species of plant to cease their growth. The winters, however, are still relatively mild when compared with those of higher latitudes, having their coldest months with mean daily minima rarely below −2°C (28°F) and usually only about three months with mean minima below 0°C (32°F). So, although trees and shrubs of the deciduous type manage to survive and indeed to dominate the biome, in the overall biological spectrum of life-forms it is in fact the *hemicryptophyte* species that make up the bulk of the flora. These are the perennial plants which die back during the winter to buds which are borne at about ground surface level. Here they are protected from the worst of the frost, often by a layer of leaf litter from the trees, but are ready to respond quickly during the early spring and take best advantage of the limited growing season.

The structure of the temperate deciduous forest is simpler than that of the tropical rain forest, having usually only two canopy layers. The main trees have their canopy at about 15–30m (50–100ft) and below this is developed a shrub and sapling layer at about 5–10m (16–33ft). This less complex structure does mean, however, that there is often more light penetrating to the floor of the forest than is found in the tropical biome. So the ground flora is well developed and diverse. Epiphytes and lianes, on the other hand, are much less common, especially among the higher plants. Some ferns can act as epiphytes, but on the whole this role is adopted by mosses and lichens; these are best developed in regions of high rainfall, where the atmosphere is constantly humid.

The deciduous canopy of trees and shrubs provides an opportunity for members of the ground flora to exploit a spring growing period when the canopy has not fully opened and the light levels at ground level are high. For this reason, many of the ground-dwelling plant species conduct most of their growth, flowering and even fruiting before the summer is well advanced. Later in the year, they are replaced by deep-shade-growing species which have highly efficient mechanisms for catching and using light at low intensity and so are able to survive even beneath a fully expanded canopy.

The productivity of the deciduous forest is very variable, ranging each year from about 0.6 to 2.5kg/sqm (0.15–0.62lb/sqft), averaging at about 1kg/sqm/yr (0.25lb/sqft/yr). Its biomass, when mature, lies in the region of 40–50kg/sqm (10–11lb/sqft).

Temperate deciduous forest is located mainly in the eastern USA, in western Europe north of the Alps up to Scandinavia, and in eastern Asia. There is considerable variation in the diversity of tree and shrub species in the different areas, however, and this could relate to their histories during and since the ice age. In North America, the advance of the glaciers from the north led to the forests retreating southward and their course was not impeded by any mountain ranges, since the main mountains—the Rockies and the Appalachians—run roughly north–south. In Europe, however, the forests retreating from the north met up with the Alpine and Pyrenean glaciers developing in southern Europe and cutting off their retreat. Many of the tree species were increasingly restricted in their distributions until they eventually reached the point of extinction under the stress of this pincer movement of ice. The Chinese forests are also richer than the European ones—they too escaped the worst effects of the ice age. Perhaps the eastern Asian and European forests once formed a continuous belt, for there are some tree groups which they have in common. An interesting tree of this type is the wing nut. There are now eight species of this tree in the world, seven of which are in China and Japan and the eighth in the Caspian outlier of the temperate deciduous forest biome in Iran. Before the ice age this last species was widespread through Europe, as is shown by finds of its pollen in a fossil state, but it is one of the trees which almost became an ice casualty in the West, although surviving as a relict in the Iranian forests. It also provides a link with the equivalent forests in the Far East.

In contrast to tropical forests, the canopy of deciduous forests harbors a very limited number of mammals. The absence of a complex series of layers, coupled with the highly seasonal nature of the canopy vegetation, precludes the development of specialized, restricted niches, especially in those species that are permanent residents. Generalists are more the order of the day. Monkeys, a dominant arboreal group in the tropics, are absent from the temperate deciduous forests of Europe and North America, but there is one species in the deciduous forests of Japan. The Japanese macaque lives primarily on the ground and has a thick coat to resist the cold. During the winter Japanese macaques search for food under the snow. They display great intelligence in getting food and keeping warm during the winter, even to the extent of immersing themselves in the water from hot springs.

Trees in this biome tend to produce nuts and winged seeds rather than fruits, so the fruit- and seed-eating niche is essentially absent. Some fruits are to be found (for example, apple, rose, hawthorn, blackcurrant etc) but they tend to come all at one time and so form a part of the diet of more generalist herbivores.

Throughout the summer, productivity builds up to a crescendo, and this bountiful harvest at the end of the summer is all-important in the survival of the forest's mammal and bird fauna. They follow one of several different strategies to survive the cold and impoverished winter. Fall is a time of feverish feeding and collection.

Many species, unable to escape the winter, sleep through it in protected dens. The body temperature of true hibernators drops and their lowered metabolism is sustained by their fat stores. Hedgehogs and dormice adopt this strategy and sleep profoundly. Squirrels, bears and badgers do not lower their body temperature and instead go into a torpor during the winter which is punctuated by wakeful periods. Squirrels feed during these periods on caches of nuts that they stored away during the fall. Animals like squirrels may, by their seed- and nut-gathering activities, have a considerable effect on the distribution and survival of deciduous forest trees.

Some species lay down fat stores in their bodies to enable them to make long migrations to their winter quarters in the south. Many birds adopt this strategy, especially those such as the warblers that feed predominantly on insects. Willow

▼ **Frozen monkey**—the Japanese macaque (*Macaca fuscata*) is the only species of monkey to occupy temperate deciduous forests, where it forages on the ground during winter.

▲ **Forager on the woodland floor.** The Eurasian badger (*Meles meles*) typically searches leaf litter for earthworms, mice, voles, snails, insects and fallen fruits. It is chiefly nocturnal, spending the day in complex burrows called setts.

► **Fighting for dominance.** Two American Red foxes (*Vulpes vulpes*) engage in a ritual battle during the fall in a North American woodland. The one on the right is the likely winner of the encounter as he is standing the highest.

◄ **Spangle galls on oak leaves.** The Spangle gall wasp (*Neuroterus quercus-baccarum*) induces the development of small flattened galls on the underside of oak leaves. The wasps develop in these galls, and for them to survive the winter the galls have to fall off the leaf before the leaf falls. Those still on the leaves are left exposed to the cold and perish.

warblers which spend the summer in English woodlands make a 9,500km (6,000mi) journey to Africa, south of the Sahara, and then return the following year to exactly the same place in the same English wood.

Animals that neither escape nor hibernate during the winter have to make the best of it, eking out their fat reserves laid down from food gathered in the fall with whatever can be found. Deer and wild pigs grub around for roots and vegetation, and browse on twigs and bark. Birds feed on buds and berries remaining on the trees. Resident insectivores search the litter layers for torpid insects and earthworms, while tits search the branches for overwintering insects and their eggs.

The other characteristic feature of the fall in temperate deciduous forest is of course leaf fall. The forest floor becomes carpeted with a thick layer of dead leaves which may be built up into drifts by the wind. Beneath this thick leaf layer is a favorite hibernating site for hedgehogs. This litter layer persists during the winter because the temperatures are too low for the decomposers to make much of an impact. Litter undoubtedly provides an important insulation blanket against the winter cold for both plants and animals.

Although it is natural to concentrate on the larger and more evident animals of any ecosystem, they are in fact supported by only a very small proportion of the total annual primary production. About 60–70 percent of the annual net primary production of temperate deciduous forests is stored in the form of woody tissue by the trees. Of the remainder a small proportion supports the grazing foodweb (only a few percent of the total) and the rest passes to the decomposers. When the tree dies the accumulated energy in the woody tissues also goes to the decomposers.

The deposition of leaves comes at a time when the temperatures are falling, so that little apparent change occurs to the litter during the winter. The rains compact and soak it, and bacteria and fungi invade the leaves. In the spring, as the temperatures increase, the decomposers become more active and abundant. The plant material deposited the previous fall is consumed by a succession of invertebrates which break it up (comminute it) into fine particles. Larger species, such as earthworms and woodlice, chew up the leaves first and extract some energy for their own needs, often materials that have already been produced by the action of fungi and bacteria. The feces of the earthworms and woodlice are still rich in energy and nutrients, and are attacked by more bacteria and fungi and consumed by smaller decomposers such as springtails and dipteran larvae. Their feces are eaten by mites, and in turn the mite feces are consumed by protozoa. At each step the plant material is ground down and digested a little more until it has been reduced to soluble nutrients that are washed into the soil for use by the forest plants once again. Each of these groups of different-sized comminuters has its own group of predators, again appropriately sized. The mites are attacked by predatory mites and pseudoscorpions, while the springtails are eaten by small carabid beetles and spiders. The earthworms and woodlice are prey to bigger carnivores such as moles, shrews and woodland birds as well as to large carabid beetles. Whereas the grazing foodweb of the forest is fairly evident and easy to understand, decomposer foodwebs are a tangle of different-sized, coexisting and interrelated foodwebs in a medium that is difficult to look at and sample. It is therefore little wonder that our knowledge of decomposer systems is lagging behind that of the grazer systems. Leaves of different trees are processed at different rates by the decomposers. Oak litter takes more than one year to be broken down, so that a thick litter develops with the pattern of comminution clearly visible throughout the layers of the litter.

In the spring and summer animal activity accelerates. Migrant birds return from their winter quarters in warmer

climates and the males set up territories and begin to sing. These bird songs serve not only to attract mates, but also to proclaim that the territory is occupied. This is the time when fairly accurate censuses of breeding populations of birds can be made. Experienced ornithologists can identify different bird songs and, by walking through a piece of woodland, pick out the singing male in each territory. By multiplying the number of territories of a particular species by two (male + female) the number of breeding birds can be assessed.

Many insect species that feed on tree leaves time the hatching of their eggs to coincide with the spring flush of new leaves. There are two advantages to be gained by the insects. The young leaves are rich in nitrogenous compounds which are used by the insect to make proteins, and they have not yet developed toxic compounds to prevent insect grazing. Older oak leaves have low nitrogen levels and develop increased tannin content as they mature which discourages insect herbivores.

Life in the temperate deciduous forest is one of careful timing and the prudent use and storage of resources. Without these qualities life in a seasonal biome is not possible PDM/BDT

BOREAL FOREST

Coniferous forest... Location and climate... Peatlands... Soil types... Productivity and flora... Population cycles in animals... Larger mammals... Beavers... Birds... Insects... How the Red squirrel avoids its predators...

THE boreal region is named for Boreas, Greek god of the north wind, and it is indeed characterized by a chilly, inhospitable climate.

The deciduous habit is advantageous only where the growing season is long enough to support the development of a new leaf canopy each year. Proceeding polewards, the summer season becomes increasingly short, and as a result any adaptation which permits the plant to start its photosynthesis as soon as temperatures begin to rise in the spring must be of great benefit. The evergreen leaf proves to be such a useful adaptation, but only when it is combined with the capacity to resist extreme cold in winter. The broad evergreen leaf of the tropical biomes is not found here, but the needle-like leaves of the conifers are extremely effective under these conditions. Their advantage over the broad evergreen leaf is probably their better retention of heat. They do not radiate heat as rapidly, nor do they lose much heat to the wind. In addition, the conical form of many conifers, such as spruce and fir, is efficient at shedding snow which would otherwise cause structural damage to the branches as a result of its weight.

The boreal forest is not represented at all in the Southern Hemisphere, as there is no continental land mass, apart from the southern tip of South America, which occupies the appropriate latitudes. In the Northern Hemisphere, however, the boreal forest forms a continuous belt around the entire globe over the continental regions. Over much of this range it joins onto deciduous forest on its southern boundary, except in the central part of the continents where it meets the temperate grassland. To its north lie the wastes of the tundra. Often the boundary of the boreal forest with the tundra is marked

◄▲ **Circumboreal species.** The lynx (*Felis lynx*) and the wapiti, elk or Red deer (*Cervus elaphus*) both occur in the North American and Asian/European boreal forests. The main prey of the lynx LEFT is small game. This one is from Europe. Shown ABOVE is the North American subspecies of Red deer known as the elk or wapiti, although some authorities would regard this subspecies as a different species altogether.

by a zone of birch forest in which the conifers become more abundant southward.

The overall precipitation of the boreal zone is not high, often of the order of 400–600mm (16–24in), and becomes even lower in the far north, occasionally as little as about 150mm (6in), which is similar to that of many deserts. But the very low temperatures for much of the year result in very low evaporation rates, so water is not in short supply.

Indeed, over much of the boreal zone water is in such excess that it impedes the process of decomposition, and the litter from growing plants accumulates as peat. These peatlands take a variety of forms. In the western, oceanic areas they are massive domed bogs with few trees on them, but in the more continental regions they are forested and frequently have a series of ridges and hollows forming along their lines of contour. These are called "string bogs."

Frost is a regular feature of the boreal climate; only about two months in the year are frost-free. It may be the frost which contributes to the development of patterns in the string bogs of the region. In the continental parts of the boreal biome, such as in the Siberian taiga, temperatures of −60°C (−76°F) are frequent, and it is remarkable that the trees can survive such extremes.

The soils are usually acid and lacking in fertility. Those which are well drained and have not developed into peatlands are often *podzolized*. This means that the nutrient elements become leached out of the upper layers of soil and deposited again in a hard layer lower down in the soil profile. The loss of such elements as iron from the upper part of the soil results in its becoming bleached and pale; where redeposition occurs the soil is stained red in a distinct layer. There are no earthworms present, because the leaf litter is unpalatable to them, so that these layers in the soil do not become mixed up again.

The productivity of the boreal forest is limited by the short growing season, and on average each year is only about half that of the deciduous forest of the temperate regions, at 0.5kg/sqm (0.12lb/sqft). The range, however, overlaps considerably with that of the deciduous forest, from 0.2 to 1.5kg/sqm (0.05–0.37lb/sqft) each year. The biomass, on the other hand, which ranges from 20 to 50kg/sqm (5–12lb/sqft) is essentially similar to that of the deciduous forest, the lower figures being derived from the northerly birch-conifer forests on the edge of the tundra zone.

The fact that a similar biomass of trees can attain only half the productivity of the deciduous forest shows that the efficiency of the productive process is greatly reduced under these conditions of a short growing season.

The overall diversity of the flora is lower in the boreal forest biome than in the more southerly forested biomes, and some ecologists consider that this may be a result of its relative youth. The retreat of the glaciers from the boreal region has been a relatively recent event in many areas. Whereas the temperate regions had lost most of their glaciers by about 14,000 years ago, parts of the boreal forest zone were glaciated until relatively recently and, at higher altitudes, glaciers still persist in parts of the region, as in Norway and in the northwest of North America. So the boreal forest is thus a young biome.

The relatively high biomass of the coniferous forest, together with the value of softwoods for timber and pulp, has led to its extensive exploitation for forestry in many parts of its range.

Zoological studies in this biome have made several important contributions to animal ecology generally. Trappers have found the area rich in various fur-bearing animals and one of the classic long-term studies on the population dynamics of predators and their prey derived from the 19th-century records of trappers' catches sold to the Hudson Bay Company in Canada. These records showed, for two species in particular, the Snowshoe hare and the lynx, that the numbers of trapped animals, and therefore by inference their populations, rose and fell together, and the oscillations in population size had a periodicity of about 10 years. This remarkably regular pattern has been considered by many ecologists. A variety of explanations have been proposed and challenged. The most recent theory comes from a mathematical analysis of population models: it suggests that the pattern is produced by the interaction of three things: the reproductive rates of the lynx and Snowshoe hare, the efficiency with which the lynx preys on the hare, and the time needed for the vegetation to regenerate following overexploitation during the peak in the hare population. Since Snowshoe hares have a relatively high rate of reproduction the shortage of food is probably the most important regulator of the periodicity. In the boreal forest the regularity of the cycles of population change in these two species seems to have a resulting effect on some of the other species present. The Snowshoe hares are also prey for the American Red fox and the goshawk. In years when the hares are scarce these predators have to find alternative prey in the form of grouse, especially the Willow grouse, which migrates into the boreal forest from the tundra regions to the north during the winter. This periodic predation produces 10-year population

fluctuations in the grouse, and the relative abundance of food also affects the survival of the predators, particularly in the case of the fox which cannot move so easily to a new area as can birds of prey.

Superimposed on these 10-year cycles are 4-year cycles of population fluctuation based on small mammals such as voles and lemmings. Again the periodicity seems to be determined by the same set of interacting elements, but this time with a shorter time base. Predators which customarily prey on these small rodents move to alternative prey in lean years, inducing 4-year cycles in them too. Thus in Norway the Willow grouse has a 4-year cycle of population fluctuation induced by prey switching by Arctic foxes and the Red fox.

Larger mammals of the boreal forest include the Gray wolves and their prey, the caribou, reindeer and moose. In winter, wolves hunt these herbivores in packs, often splitting into two groups to engage in a pincer attack. Alternatively one group causes a diversion while the other sneaks in close. Nonetheless they usually only take young, infirm or diseased animals rather than prime specimens. Wolves have been persecuted by farmers because livestock are easy meat to them, and they are now far less common than previously. Black bears also have ranges that extend northwards into the boreal forest.

During the winter most of these mammals stay within the forest, protected to a degree by the vegetation. Species that do not hibernate and remain active show adaptations to enable them to move more easily over snow. The reindeer and moose have splayed-out feet to spread their weight, and similar expansions of the feet can be seen in the Snowshoe hare, lynx and Willow grouse.

The American beaver is an important mammal in boreal forests, particularly near watercourses where deciduous trees such as aspen, birch and willow grow. These mammals can cause substantial changes to the forest vegetation, out of all proportion to their size. They feed on the bark of deciduous trees, and if the tree is bark-ringed it dies. However, beavers also fell trees to construct dams in woodland streams and rivers. In the resulting pond, trees are also used to construct the lodge. Both dam and lodge are splendid feats of engineering with due regard taken in the design to counteracting water pressures and overflows. Mud and vegetation are used to fill in between the tree trunks. The lodge contains the above-water beaver nest, whose entrance, however, is underwater. The beavers provision the nest chamber with wood, which sustains them during the winter. Even when the water is frozen the beavers are protected within their lodge from predators such as the wolverine. This predator is large enough to climb trees and drop onto unsuspecting, passing caribou.

Most birds migrate southward for the winter, but two species are specially adapted to cope with one of the most plentiful and nutritious foods in the boreal forest, conifer seeds. These seeds are tightly held inside the fir cones and can only be extracted by the skillful use of the modified beak of the crossbill. The Red crossbill of Europe, Scandinavia and Russia and its American counterpart the White-winged crossbill both have bills that cross at the tip.

There has been some considerable interest in the insects of these conifer forests. As man becomes more aware of the value

▲▶ **Beaver and its dam.** RIGHT a beaver pond and its dam in the North American boreal forest. ABOVE A North American beaver (*Castor canadensis*). Dams are built across streams to impound water, using mud, stones, sticks and branches. The water must be deep enough for beavers to be able to swim under winter ice from their lodges to food caches.

▼ **Equipped to crack pine cones.** The tips of the bill of the Red crossbill (*Loxia curvirostra*) are crossed—an adaptation which enables it to prize open the cone scales and pick out the seeds, as it moves sideways along branches. The female is green, the male red.

of these forests as useful natural resources for the softwood and paper industries, so those insects that damage and defoliate the trees come under close scrutiny. The large tracts of virtually single species forest are prone to attack by a variety of defoliating lepidoptera. Populations build up over several years until the densities of caterpillars are so great as to cause total removal of the needles from the conifers, which may kill the trees. The Spruce budworm causes extensive mortality in this way to several conifer species in Canada. This moth species is highly mobile as an adult and can move up to 100km (62mi) in a single generation. Birds are important predators of the caterpillars, and also at very high densities the budworm population may crash due to the outbreak of a virus disease. Managed areas are now sprayed either with virus particles or with insecticides in an attempt to control the budworm.

The Spruce budworm is only one of a number of important defoliators. The Canadian Insect Forest Survey has been monitoring population densities of forest insects for over 50 years in an attempt to look for long-term trends in the patterns of population change. The picture is very complex, with some species undergoing dramatic changes in density while others maintain low and fairly constant densities from year to year. One analysis of these data has attempted to explain these differing patterns. It showed that a lack of stability in population size from year to year was found in lepidoptera of two kinds. Either they were *polyphagous* and could feed on a range of different tree species; or they were *monophagous* species present in areas where their preferred tree was the only tree species present. In both cases the key feature is an abundance of readily accessible food. Lepidoptera whose food plants are patchily distributed among non-preferred plants maintain lower and more constant populations from year to year. The number of other competing herbivores also influences the success of a population.

As man makes greater use of these vast boreal forests, their integrity is in jeopardy. Only by acting with thoughtful regard for the boreal forest community as a whole will it remain as one of the true wildernesses on earth. PDM/BDT

Battle in the Boreal Forests

How the Red squirrel avoids its predators

Most squirrel species live out their lives either totally in trees or totally on the ground. Red squirrels, however, use both habitats, which brings them into conflict with two types of predators—aerial birds of prey and terrestrial carnivores. Red squirrels tend to avoid open areas where they are most obvious, keeping to conifer and deciduous woodland. They also possess a number of adaptations that help them avoid and escape predators. Their eyes are large and set high on the head, giving a wide angle of view. They possess long claws to help them grip, and a long bushy tail as an organ of balance—both aids for rapid jumping and climbing.

The main year-around predators of Red squirrels in the boreal forests of Scandinavia are the goshawk and Pine marten. Less important predators include owls, other hawks, buzzards, Red foxes and stoats, and to a limited extent domestic cats and dogs.

Depending on the season, Red squirrels forage in two ways. During the winter their main sources of food are conifer seeds in cones sited at branch apexes. When feeding on these, they are particularly vulnerable to daytime predation by birds of prey. In winter, therefore, foraging activity is concentrated around midday, shortening the period when they are exposed. In spring and summer most foraging occurs on the ground in thicker undergrowth, most often searching for fallen seeds. Foraging starts just before sunrise and continues to sunset, with a rest period around midday. While foraging at this time of year they are better protected from daytime predation by birds of prey, and avoid exposing themselves at night to the nocturnal Pine marten and owls.

The risk of predation increases when adults move from one feeding ground to another, and during mating in the breeding season. Juveniles are particularly vulnerable at times of play, and subadults when they are driven from their parents' home ranges in late summer and early fall.

How then do Red squirrels avoid predation at vulnerable times, and what are the predators' tactics? When feeding on spruce cones, squirrels forage at branch bases near the trunk and on pine cones within the dense foliage mass. Both these strategies prevent detection through the foliage. When detected and in obvious danger from birds, Red squirrels freeze and can remain motionless for up to half an hour. If the danger appears less imminent they produce a repeated "chucking" call, stamp their feet and flip their tails—all behaviors intended to warn other family members of the danger. When taken totally by surprise, some individuals have been known to fall several meters to lower branches to escape the attacker.

Goshawks prey heavily upon squirrels during times when the squirrel population is numerous—during January and February in boreal forests Red squirrels occupy 70 percent of the goshawks' diet. In these winter months goshawks do not hold territories, but roam widely; subadult birds are particularly wide-ranging—in one winter a subadult goshawk on a 16 hectare (40 acre) island in a Swedish lake caught 80 percent of the squirrels. Goshawks usually attack from short distances (around 50m/165ft), generally initiated from a perch in dense tree cover, or less often through low-flight attacks near ground level. It takes about 1 hour for a goshawk to devour a 280g (10oz) squirrel, after which it remains sedentary for the rest

◀▲ **Predator and its prey.** LEFT The Red squirrel (*Sciurus vulgaris*) of Europe's boreal forests is heavily preyed upon by Northern goshawks (*Accipiter gentilis*), particularly at times when the squirrel population is high. The Northern goshawk, in addition to northern Europe, also occurs throughout Asia and the Americas. The one shown ABOVE is an immature individual from the American boreal forest; it has a Gray squirrel (*Sciurus carolinensis*) in its talons.

of the day. The next day, the goshawk is active for only 5 percent of the daylight hours, increasing to 11 percent on the third day. Most of their time is spent perched watching for squirrels.

When hunted by Pine martens, Red squirrels adopt a different strategy—they utter a shrill scream and climb to the outer branches that are too thin to bear the marten. Pine martens are efficient hunters and take up to 10 percent of the Red squirrel population in the winter—on average they eat eight squirrels each month. As well as hunting in the open, martens will enter dreys to find their quarry—sometimes leaving the carcasses dangling from the drey entrance as macabre evidence of their meal. Each hunt involves traveling about 5km (3mi) and squirrel dreys may be used for sleeping overnight. In winter they climb few trees (0.8 times per 10km; 1.3 times per 10mi), preferring to search on the ground. Both squirrels and martens prefer mixed conifer forests and older woodlands, and up to 71 percent of Pine martens eat Red squirrels, particularly at high squirrel densities. When squirrel populations are low the number of martens taking squirrels drops to 42 percent, while the number of Bank voles taken increases. OG

ARID SCRUB

Typical vegetation... Extent and productivity... California... Southern Australia... Chile... Mediterranean Basin...

THE scrub biome is characterized by low, woody perennial plants which have a *sclerophyllous* habit. This means that the leaves are small and leathery, and thus well adapted to withstanding periodic drought. Such vegetation is most common in those regions which have a mediterranean type of climate, that is one with hot, dry summers and mild, wet winters. The total annual rainfall is often in the range of 250–500mm (10–20in), and this falls almost entirely during the winter months. The monthly average temperature in summer is often in excess of 20°C (68°F), and in most areas of this biome frost in winter is rare.

Although the most obvious plants are the shrubs, an examination of the life-form spectrum shows that the commonest type of plant is usually the therophyte, or annual. In a climatic regime with severe summer drought the ability to remain in a dormant state through the hot summer is a great advantage. Many of these annual species are actually stimulated into germination and growth by the lower temperatures of the fall, which contrasts strongly with the warmth-stimulated seeds of the temperate regions. But fall germination means that a mild, moist period lies ahead during which establishment can take place. Then they flower and fruit in the

Sagebrush Succession

The Californian sagebrush is rich in volatile chemicals which deter grazers and inhibit the germination and growth of plant competitors; also its growth in an area of scrub presents a fire hazard as a consequence of the inflammable nature of these *terpenoid* substances. On average, fire sweeps through the Californian chaparral every 25 years, after which there is a period of gradual recovery.

Immediately after fire, the volatile terpenoids are all burned off, as are the shrubs which act as their source. This means that all the more delicate and sensitive species which were being suppressed by the chemical warfare of such plants as the sagebrush are now able to develop and thrive. But within three to four years the sagebrush shrubs have begun to invade once again.

Over the next few years, the shrubs grow and produce a complex of patches in the vegetation, for where they expand their canopies the leaching of terpenoids from their leaves suppresses growth and produces bare areas, devoid of other plant life. Other species survive only in the clearings between the sagebrush. The soils become more and more rich in terpenoids and are covered with leaf litter from the inflammable shrubs.

Eventually, in about 25 years, the scrub has grown so tall and dense that fire becomes inevitable. It rages with great ferocity and leaves a bare but terpenoid-free environment in which the cycle begins once again. There is one great advantage with this strategy from the shrub's point of view. The fire destroys any longer-lived, more robust species which might otherwise replace it. So its inflammable chemicals protect it in a number of different ways. PDM

► **Animals of the Californian chaparral.** (1) Collared peccary (*Tayassu tajacu*). (2) White-tailed antelope-squirrel (*Ammospermaphilus leucurus*). (3) Merriam's kangaroo-rat (*Dipodomys merriomi*). (4) Sidewinder (*Crotalus cerastes*). (5) Western collared lizard (*Crotaphytus collaris*). (6) Road runner (*Geococcyx californianus*). (7) Antelope jackrabbit (*Lepus alleni*). (8) Puma (*Felis concolor*).

8

7

warmth of spring, and return to the soil as seeds for the summer. An alternative strategy is adopted by the geophytes, or bulb plants, which also abound in this biome. These are perennials which retreat from the heat of summer into their underground bulbs and corms.

The shrubs themselves may be evergreen, in which case the leaves become very water-deficient during the summer, or they may lose their leaves in direct response to drought. Many are rich in volatile, strong-smelling compounds, the function of which is either to deter insect feeders or to compete with other plants by suppressing their germination and growth.

The main regions of the world in which this biome is found are the Mediterranean, California, Chile, the southern tip of South Africa and parts of southern Australia. In some of these regions the vegetation has been degraded by intensive farming, particularly high-intensity grazing. This often means that the full potential biomass is not attained and a low, stunted scrub is produced in its place. Under productive conditions a biomass of 25kg/sqm (6lb/sqft) can be achieved, but where the vegetation is degraded by sheep or goats, or where drought is particularly severe as in the regions bordering on desert, the biomass may be less than 5kg/sqm (1.2lb/sqft). Productivity is similarly variable, and is limited largely by the degree of summer drought. Average annual values are around 0.8kg/sqm (0.2lb/sqft), but in the really arid areas it may be as low as 0.05kg/sqm (0.012lb/sqft).

The Mediterranean basin itself is somewhat depleted of

5

6

4

animal life, owing to the long history of man's activity in the area, but in other parts of the world this biome supports a diverse fauna.

The chaparral of California is an area of thorny scrub rich in birds and other vertebrates, especially in the wet season, though during the long, hot summer many birds and some of the large herbivores move to more fertile habitats. Chaparral mammals include the herbivorous ground squirrels and kangaroo rats, both of which store seeds in their burrows.

These seeds may assist in water conservation by absorbing water vapor expired by these small mammals while they are in the burrows. Larger species include the Collared peccary, a small relative of the pigs, which is omnivorous, the common Antelope jackrabbit, which as its name suggests is an accomplished runner, the widespread Mule deer, and now less commonly wolves, Grizzly bears and Mountain lions. Among the birds, the roadrunner or Chaparral cock is somewhat unusual. It belongs to the cuckoo family but is non-parasitic. It is a poor flier but can run fast. Reptiles and rodents are the roadrunner's principal food. The centerpiece of the Thanksgiving dinner, the domestic turkey, came originally from the open scrubby forests of Mexico and California. It was taken to Europe by the Spanish and subsequently reintroduced to North America in its domesticated form.

In southern Australia the semi-arid scrub is called mallee. Bird life is very varied. As might be expected from such a seasonal habitat, fruit feeders are uncommon, while the commonest lifestyle is that of the seed feeders. One such is the Mallee fowl. This bird does not incubate its eggs by sitting on them but instead builds a compost heap and lays the eggs on it. The male then controls the eggs' temperature by adding compost to, or removing it from, the heap. The carnivorous lifestyle is also common, there being several species of falcons, goshawks, owls and butcher birds characteristic of the mallee.

In South America the matorral of Chile has a mediterranean climate. Small mammals of this area include the degu, a rat-sized rodent with sharp claws which it uses to dig for roots and tubers. The degu also selectively grazes and produces an interesting pattern in the distribution of the vegetation. Around the degus' burrows a competitively inferior herb, the sneezeweed, grows profusely, although elsewhere grasses successfully compete against it. By selectively grazing the grasses around its burrow the degu enables the sneezeweed to persist in the matorral. The shrubs and trees are not browsed by vertebrates nowadays, although guanacos may have been in the matorral in earlier times.

In the Mediterranean basin the sparse garrigue and denser maquis vegetation is under considerable assault by man's domesticated animals. Goats in particular are favored because they are very cosmopolitan feeders, doing well on poor-quality vegetation, and provide milk, wool, meat and skins. They manage with little or no water, can climb trees to get at what they browse on, and show ingenuity in finding food when it is scarce. The special problem in all these arid scrub lands is that in the dry season there is nothing to protect the plants from being grazed. Farmers do not lay up fodder stocks for this period, and so the animals keep on grazing although the plants are not able to grow. This results in an impoverished and sparse vegetation. The vegetation is further modified by fire, for in the dry season it is tinder-dry and burns easily, especially because of its volatile oil content. This favors those plant species that are stimulated by fire or can tolerate it. PDM/BDT

► **Tending his incubation mound.** This male Mallee fowl (*Leipoa ocellata*) may work on his nest in most months of the year, digging out a hole, then scooping leaf litter into it and covering it with sandy soil. Once laid, the eggs are ignored by both parents, as are the chicks.

DESERTS

Climatic conditions... Productivity... Soil structure... Animal life, including mammals, birds, reptiles, amphibians and invertebrates... How animals survive...

THE most characteristic feature of desert is drought. The lack of available water is the overriding ecological factor affecting both plant and animal life, and this lack may be due to low rainfall, high evaporation rates, or both. There are some deserts, such as the Gobi Desert, which are not hot, but are protected by high mountain ranges from the influence of rain-bearing air masses. But most deserts are hot, being in tropical or subtropical latitudes where pressure systems are high and rainfall is correspondingly low. Extremely arid deserts may have a rainfall of less than 60–100mm (2.4–4in) per annum (the term "arid" cannot be precisely defined in rainfall terms, because much also depends upon evaporation rates, which in turn depend on temperatures). In these extremely arid deserts, vegetation may be absent or confined to the wadis (dry river beds) which carry surface runoff water after rain.

Arid regions may have rainfall up to 150–250mm (6–10in) per annum, and these have a more diffuse cover of vegetation; while semi-arid areas have up to 250–500mm (10–20in), a grass or scrub vegetation, and may support some arable farming. The rainfall which arrives may come largely in winter (as in the Middle East) or in summer (as in the southern Sahara and central Australia). The Namib Desert in southwestern Africa receives most of its precipitation as fog.

Daytime air temperatures in the hot deserts may rise to over 50°C (120°F) and the surface sand temperature may be as high as 90°C (195°F). During the night these temperatures fall again and the relative humidity rises. Below the surface of the soil, for example in the subterranean burrows of rodents, the high temperatures and low humidities are not experienced, so life in a burrow is considerably less extreme.

The sparse vegetation of deserts means that the above-ground plant biomass is low, varying between 0 and 5kg/sqm (0–1.3lb/sqft, the latter value being for semi-arid scrub). Since many desert plants have extensive root systems, the below-ground biomass may be considerably greater, sometimes reaching 24kg/sqm (6lb/sqft), which means that much of the energy available to consumers and decomposers lies beneath the surface of the soil. The productivity (above ground) of deserts is also very low, rarely exceeding 250g/sqm/yr (0.06lb/sqft) even in scrub habitats. When it is considered that about 14 percent of the world's population lives in these dry areas, relying largely on pastoralism, it is not surprising that overgrazing often results, leading to a further reduction in plant biomass and the spread of desert conditions (*desertification*).

The plant life of the deserts consists in part of perennial species which are capable of survival under dry conditions. They are able to reduce water loss during times of stress, but the side effect of this is that their photosynthetic rates are also kept low. Among these perennials are the New World cacti, which have swollen stems and reduced leaves and branches, their entire structure tending towards the spherical, whereby they radiate heat most effectively and reduce the surface area available for water loss. Their ecological equivalents in the Old World are the euphorbias, which have evolved along similar structural lines. Woody shrub plants are also frequent in deserts, and these are usually deciduous during the dry periods. An alternative strategy is that of the therophyte, which spends most of the year as a dormant seed and then germinates when conditions become suitable, that is after periods of rainfall.

The soils of deserts generally lack structure, but the particle size of the upper layers is critical in determining the evaporation losses and the rate of water penetration. A fine-particle soil has poor rain penetration and therefore high evaporation loss, whereas a sandy or rocky soil has good penetration and less evaporation.

Since much of the plant biomass is below ground, the bulk of the nutrient cycling of the desert ecosystem also takes place in the soil. Dead root tissue is immediately available to the microbial components of the system, and the release of elements thus occurs within the soil. Added to this is the litter which accumulates on the soil surface, particularly from the leaves of deciduous scrub.

A further feature of desert soils is the high content of salts and lime, including sodium chloride, calcium carbonate and

Locomotion on Sand

Animals which live in sandy desert conditions show some special adaptations to enable them to move more efficiently over the shifting surface. Although camels have only two toes per foot, each toe is very broad and prevents the camel sinking into the sand. The fringe-toed lizard, as its name suggests, has its foot area enlarged with fringes of scales, while some geckos ABOVE have webbed feet which not only prevent sinking but are also useful in digging. Gerbils, jerboas and kangaroo rats, as well as kangaroos proper, cover the shifting substrate in large bounds. Gerbils have tufts of hair on their feet, which gives them a good grip on the sand. The hair may also insulate them from the hot surface layers of sand. The sand lizard shuffles its feet as it walks forwards. This serves to push aside the hot surface of sand grains and expose the cooler layers beneath.

Perhaps the most complex adaptation to moving on sand is that found in desert vipers and rattlesnakes. These reptiles move by a process called sidewinding in which sections of the snake's body are moved sideways in a hump. The movement is very fluid and fast, with waves of these sideways humps passing down the length of the snake.

◄ **Alert to possible danger,** a Cape ground squirrel (*Xerus inaurus*) stands upright while eating some seeds. During the hottest part of the day these squirrels remain underground.

◄ **Web-footed lizard** OPPOSITE—a gecko (*Palmatogecko vangei*) in the Namib Desert (see box).

▼ **Nymph and cactus.** A nymph of the grasshopper *Taenipoda auricornis* eating the fruit of the cactus *Melocactus elessertianus* in a Mexican desert. Cacti have several adaptations to prevent water loss and to store it; this provides an important water and food source for many New World desert animals.

calcium sulfate (gypsum). Owing to the evaporation of water from the soil, many of these chemical compounds become concentrated in the surface layers and may even crystallize there, making it very difficult for any plants except salt-tolerant ones (*halophytes*) to survive. This process of salination becomes even more acute if man raises the water table in soils, bringing the salts nearer the surface, as when he irrigates his crops.

Despite their harsh environment the deserts are a home for a surprising range of animals. Large mammalian herbivores such as camels, sheep, goats and donkeys are familiar elements of the desert scene. These species have been domesticated by the wandering desert tribesmen and are well adapted to the hot, dry desert conditions. In addition several non-domestic species, gazelles and ibex for example, also compete for the sparse grazing.

Small mammals such as jirds, gerbils, kangaroo rats, spiny mice and ground squirrels are common in deserts. In one study in the Negev Desert in the Middle East at least one species of mouse was to be found in every 2.6sqkm (1sqmi) of the whole desert. As might be expected, this variety of small mammals provides a useful food source for a range of carnivores. Wild cats, caracals, several species of fox and hyenas are the main

predators, and in Australia feral cats (domestic cats which have "gone wild") have been found, apparently thriving, deep in the desert regions.

Birds generally have higher body temperatures than mammals, and so the heat of desert localities is not so much of a problem. Consequently the desert has a rich but fairly undistinguished avian fauna. The ability to fly enables birds to seek out available water sources over vast tracts of arid desert. Large flocks of birds congregate in the mornings at these watering places before dispersing to look for food. Perhaps the most interesting desert birds are the flightless species such as the ostrich, which inhabits the desert regions of Africa. Its range is more limited than that of other desert birds, since without the ability to fly it cannot move far from the watering places. In the Australian deserts the emu has a similar lifestyle.

Reptiles are well represented by a variety of lizards and snakes, including the infamous Gila monster and the rattlesnakes. They are generally carnivores and feed on insects and small mammals.

It is somewhat surprising that deserts, noted for their lack of water, contain some amphibians and fish, both of which need water at some stage of their life. The young amphibians (tadpoles), develop very rapidly in small temporary rainwater pools or in burrows which are flooded during the rains. Small streams may be present in deserts, starting as springs and flowing a short distance before becoming totally absorbed by the porous desert soil. These small streams are hot and frequently salty. Such conditions are tolerated by several species of pupfish which are highly adapted to these hostile habitats. A variety of invertebrate animals (crustaceans, rotifers and insects) are also found in these desert aquatic habitats.

Terrestrial arthropods are common. Many are strictly nocturnal, such as scorpions, centipedes, spiders and woodlice. They rapidly retreat under stones or down burrows as the sun rises. Others are abroad during the daytime, including the black tenebrionid beetles and the velvet mites. These mites are toxic to predators and have been known to kill mice: nonetheless they are collected by the people living in the Sind Desert in Pakistan to use in their cooking.

To be able to survive in the desert, animals must cope with the twin problems of the heat and the dryness of the habitat. They have to maintain their body temperatures within acceptable limits. "Warm-blooded animals" hold their body temperatures constant, while "cold-blooded animals" tend to have a body temperature close to that of their surroundings. By moving into cooler or warmer situations, however, they can exert control of their internal temperature. Water, which is important in maintaining the internal body fluids at their correct concentrations, is lost during breathing, sweating and excretion. Since these losses cannot be entirely prevented, animals must be able to obtain water from somewhere. These two major problems of desert life are solved in desert animals by a combination of structural, physiological and behavioral methods.

◂ **Blue-tongued threat.** Most lizards attempt to escape when under attack, but the slow and clumsy Shingle-backed lizard (*Trachydosaurus rugosus*) of Australian deserts threatens predators with a gaping mouth, strong jaws, bright blue tongue and bouts of hisses.

In the large herbivorous mammals the methods of dealing with the desert's problems are mainly structural and physiological. Camels, sheep, goats and donkeys have particularly thick coats on their backs. The outer surface of this coat of hair sometimes reaches temperatures in excess of 70°C (158°F)—water at this temperature is scalding—but at the skin surface the temperature is only about 40°C (105°F). Thick coats are normally associated with keeping heat in, but they work equally well in reverse to keep heat out.

Large size itself is advantageous in desert situations. The larger the mass of an animal, the slower it will warm up in hot conditions, in much the same way that a full kettle of water takes longer to heat up than a half-full one. This may well be why the ostrich and the emu are so successful as desert inhabitants. Despite their thick coats and large size these animals still need to lose heat. One way is to radiate away excess heat to cooler regions of the habitat. The hair on the underside of the camel and donkey is much less thick, and the surface skin there is well supplied with blood vessels. These animals can therefore unload some of their excess heat into the cooler, shaded air beneath them. The ostrich uses the poorly feathered regions under its wings for the same purpose.

The metabolism of warm-blooded animals is very finely tuned, operating at a very precise temperature which is maintained by the animal. In most animals changes in the body temperature lead to abnormal metabolic activity and can threaten life itself. In humans a raised temperature or fever is considered dangerous if the temperature rises more than a couple of degrees, and steps are taken to reduce it to normal. Some desert animals allow their temperatures to drift upward during the day, by as much as 6C° (11F°) in the camel, without any danger. The temperature returns to normal during the relatively cool nighttime.

Animals usually sweat to cool down. This is a very effective method, but wastes valuable water. Sweating in desert animals is therefore kept to a minimum by some of the methods already mentioned, but some are tolerant of losing large amounts of water. Usually as the water content of a "normal" animal drops, for example by sweating, the water comes from all parts of the body. In particular, water is lost from the blood, which becomes thicker as its volume decreases. This puts a severe strain on the heart. Both the camel and the donkey are able to lose between a fourth and a third of their body weight as water, and they are able to do this without their blood volume altering. The water is lost from less vital areas. Furthermore, these species are able to drink large quantities of water extremely rapidly when it is available, and so quickly replace their losses. Kangaroos, the large herbivore equivalents of

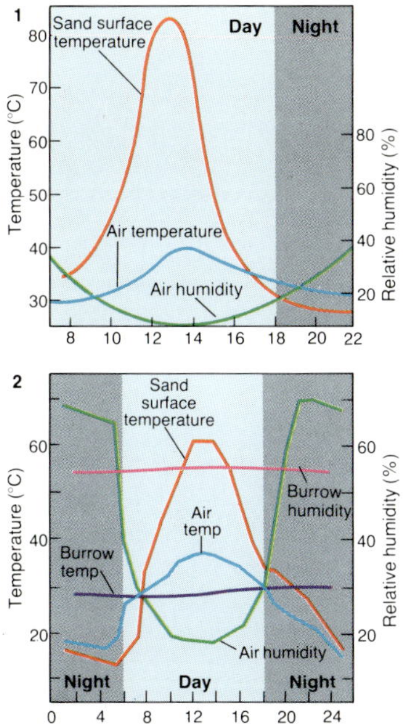

▸ **Hot and dry**—the desert environment. (1) Graphs showing the temperature and relative humidity in the Sudan desert during the day and night. (2) Comparison of environmental conditions inside and outside the burrow of a Desert fat jird (*Psammomys obesus*). Note that conditions remain stable in the burrow and fluctuate widely outside.

◂ **Golden burrower of Australian deserts.** Head of the weird Marsupial mole (*Notoryctes typhlops*), the only Australian mammal specialized for a burrowing life. Its hands are flattened with greatly enlarged claws for excavating the soil. It has no functional eyes and lacks external ear flaps.

▾▸ **Mind your feet!** Desert vipers lie in wait for their prey with only their eyes visible above the surface RIGHT. These snakes have saw-like edges to the scales on their flanks. As the snake wriggles into the sand BELOW these shovel the sand aside and over the snake's back. This species is the Common sand viper (*Cerastes vipera*), here photographed in the Negev Desert in Israel.

camels in the Australian deserts, smear saliva on their bellies, feet and tail to supplement sweat cooling.

Large animals cannot usually escape from the full heat of the sun, but almost all the smaller animals avoid prolonged exposure either by burrowing or coming out only at night. Most of the small rodents are active only during the night when the temperatures are cooler. During the day they remain in burrows where the moisture content is high and the temperatures are low—typically 25–35°C (77–95°F). Some of these small seed-eating rodents, such as kangaroo rats, can survive without drinking at all. Staying in the burrow reduces the need to sweat, but they still lose water by breathing and in urine production. The kidneys of these rodents are able to conserve some water by producing very concentrated urine. Kangaroo rats obtain all the water they need from their food—dry seeds. Seeds stored in the burrows act as edible sponges by picking up some of the water vapor in the atmosphere. Also, the bulk of the seed is composed of carbohydrates. When these are metabolized they give energy, carbon dioxide and water: this water replaces the need to drink.

Some rodents, such as ground squirrels, emerge during the day for short periods of time, but because of their small mass they quickly heat up in the sun's rays. After about 15 minutes' exposure they have to return to their burrows, where they cool down by pressing their bellies on the cool walls of the tunnel.

The carnivore's lifestyle enables it to succeed in deserts without being very specialized. Desert carnivores are usually nocturnal and use burrows or caves to keep cool during the day. Their food contains a lot of water, and so they need to drink only seldom. The carnivores therefore show few adaptations to desert life, and this is probably why feral cats have been so successful in Australia.

Although snakes are nocturnal, the other reptilian group, the lizards, are often daytime hunters. They remain in burrows or buried in the sand during the night and emerge in the morning to warm up. Reptiles are cold-blooded and need to be warm to become very active. Lizards warm up by basking, and then by moving from sun to shadow they can accurately adjust their internal temperature.

Tenebrionid beetles feed during the day. At night, they have an intriguing method of obtaining drinking water. As the temperature drops at night what little water there is in the air begins to condense as dew. These beetles stand with their abdomen upward and head downward so that the dew can condense on their bodies. The water droplets run down to the head where they are drunk.

One interesting feature of the reptilian and arthropod desert carnivores is that they tend to have very potent venoms. Where food is scarce, a predator must be able to deal speedily with any suitable prey that comes along. Scorpions and spiders of desert regions are among the most dangerous in their groups, as are many of the snakes. The only poisonous lizards are also desert dwellers. These venoms also provide a good deal of protection should anything choose to attack them. Other species use toxic materials solely for defense. Animals as different as toads, millipedes, mites, blister beetles and whip scorpions all have secretions or sprays to protect themselves against predators. PDM/BDT

SAVANNA GRASSLAND

Climate... Grasses and trees... The effect of fire... Productivity... Herbivores in relation to the plant life... Predators... Insect life... Exploitation by man...

THE savanna, with its wide, grass-covered plains, is one of the most attractive tropical landscapes. Uniform grassland with scattered trees is the vegetation typical of these tropical regions, in which rainfall is more or less confined to a summer wet season while for the rest of the year a hot, arid climate prevails. These tropical park-grasslands are found on both sides of the Equator, outside the rain-forest belt. They are best developed in Africa and South America south of the Amazon forests, but savanna grasslands also occur over much of India (in uncultivated areas), in parts of Southeast Asia, and in northern areas of Australia.

The vegetation of the savanna is dominated by one family of plants—the grasses. These are highly-evolved flowering plants in which the main growing parts are not at the stem tips, but at the bases of the leaves. This means that as the leaves are eaten by herbivores, they simply regrow and replace the lost parts. The grasses are often very tall, like the African Napier grass, which can grow to several meters in height, and many of them have developed a specialized form of photosynthesis which is particularly efficient under high light intensity and high temperature conditions. It is also a system which allows more carbon dioxide to be fixed per unit of water used than conventional photosynthesis.

The trees of the savanna also need to be efficient at water conservation because of the long dry season, so many have a distinctive water-retaining structure. Most remarkable of these is the baobab tree, which has a massive water-holding trunk. It is claimed that it can hold up to 120,000 liters (32,000 gallons). The other type of tree which is most abundant in Africa is the acacia, and this large genus of plants (about 700–800 species) is found throughout the savannas of the world, from South America to Australia. A major savanna tree in Australia is the eucalyptus, which is particularly adept at surviving fire because of the resistant buds within its bark which commence growth after the fire has passed.

Fire is a regular occurrence in the savanna biome, and this may restrict the entry of more trees into the ecosystem. The grasses regenerate well after fire, but many trees are not as resilient and fail to survive. The other factor which may limit tree invasion is the poverty of the soil; the soils are often too infertile to permit the growth of nutrient-demanding trees. It has been suggested that the trees which do survive in the savanna build up the local supply of nutrients in the soil, which then permits more demanding species to invade. But the frequency of fires will probably prevent this even when the soil becomes rich enough. Some of the grasses, including Napier grass, produce chemicals in their leaf litter which inhibit the bacteria in the soil and prevent nitrogen from being cycled and made available to the trees. In this way, by keeping the soils poor in nutrients, the grasses may ensure their own survival.

The nature of the life-forms in this biome results in a relatively small biomass of maximum value about 5kg/sqm (1.3lb/sqft). The productivity, however, is quite high, with annual values ranging from 0.2 to 2kg/sqm (0.05–0.5lb/sqft). So the productivity of the tropical grasslands is about the same as that of the boreal forest, but it achieves this with only about one-tenth of the biomass. It must, therefore, be regarded as quite an efficient biome.

Perhaps it is this productive efficiency which results in the richness of this biome in large grazing animals, and which provides a habitat in which man himself evolved during the last two to three million years.

In many parts of the tropical world, including East Africa and Australia, the savanna grasslands were more extensive in the past, particularly in those regions now occupied by tropical rain forest. When the high latitudes were experiencing glacial advances, much of the tropics were becoming more arid, and savanna vegetation was very widespread as a result. These ideas are confirmed by studies of lake levels in Africa, which were very much lower during the high-latitude glacials. The same is true of northern Australia and probably of the South American savannas too, though there is currently less information available from that continent.

The savanna fauna is characterized by a great abundance and diversity of grazing species. The East African savanna supports the greatest diversity of herbivores in the world, with over 40 species of large grazing mammals—and this in a habitat that does not appear to provide a very great number of feeding niches. In other situations the diversity and biomass are far less, particularly in areas where man has hunted excessively as in other parts of Africa and the chacos of South America.

The average vertebrate biomass on the African savannas varies from about 1,000kg/sqkm (5,700lb/sqmi) to as much as 30,000kg/sqkm (170,000lb/sqmi), although these highest values are of heavily overgrazed habitats dominated by elephants and hippopotamus. The highest biomass values for non-overgrazed savanna are around 12,000kg/sqkm (68,000lb/sqmi). These large herbivores are not spread evenly, but tend to move around in herds seeking new vegetation and waterholes. During the dry season these herd movements become spectacular migrations, as for example in the case of

◄▼ **Flushed with greenery and flowers.** The savanna grasslands of Africa are often seen as burnt and dried out. However, this is only during the dry season. When the rains come the vegetation grows rapidly and the animals can disperse farther away from their water holes. Shown here are LEFT a lioness (*Panthera leo*) resting among a bed of purple flowers and BELOW a herd of African elephants (*Loxodonta africana*) grazing.

the wildebeest, where herds of more than 10,000 individuals have been recorded moving from one site to the next.

In the African savannas the large number of different herbivore species divide up the available resources in a very specific and interrelated way. Some species, such as the African elephant, are broad-ranging generalist grazers. They take a great range of different plant species, and which they eat seems to depend principally on availability. Elephants may cause considerable modification to the habitat, particularly by uprooting trees and shrubs, and thus help to open up the savanna and maintain its grass-dominated nature. This destruction of the tree and shrub layers is something of a problem in national parks where the elephants are protected from hunting. In such conditions the elephant herds increase in size until overgrazing takes place. The Common eland is also a generalist browser and grazer.

Other species feed either in a succession or on very specific plants. Three of the most numerous herbivores feed successionally on the grasses. The Plains zebra grazes the grasses first;

it is followed by the Brindled gnu (Blue wildebeest), which cuts down the grass sward further; and finally Thomson's gazelle feeds on the short grass and also supplements its diet by browsing on low acacia bushes. Thus herds of these three species tend to follow one another across the savanna. The small size of Kirk's dikdik enables it to feed very selectively on the small nutrient-rich growth of young leaves on bushes. In areas where the giraffe is common, the trees are often characterized by a distinct browse line above which the giraffes cannot reach.

These large numbers of herbivores have a considerable effect on both the vegetation and other animals. Mention has already been made of the destructive effects of elephants, but the increase in grassland they produce favors the grazers like zebras. Feeding by herbivores is sometimes very beneficial. At the end of the rainy season in the Serengeti National Park the huge wildebeest herds move onto the so-called short grass community and graze it heavily, removing 85 percent of the plant biomass in a few days. In a comparative study of areas grazed like this by wildebeest and other areas protected from grazing, it was found that the heavy cropping stimulated new grass growth which was subsequently eaten by gazelles, while the grass in the ungrazed areas declined in quality and biomass. This grazing revitalizes the grassland and further enhances the links between the wildebeest and gazelles.

There are also examples of mutual benefit between trees and their browsers. The Umbrella tree has large seed pods that are a favored food for several browsers, including eland, elephant and dikdik. The seeds germinate far more successfully after having passed through the gut of a herbivore than if they are dropped on the ground. Although the gut enzymes may weaken or condition the seed coat, a well-established phenomenon in other situations, it is thought that the prime effect is to kill the developing larvae of a seed-eating bruchid beetle. So the herbivore gains by feeding on the nutritious seed pod, and the plant has its seeds protected by the insecticidal properties of the gut enzymes. They are also "planted" in a rich compost.

These large assemblages of herbivores are the prey for a range of predators. Lions tend to shelter in the shade of trees and venture forth on short chases to capture food. To be successful the lions have to stalk their prey and get close before revealing themselves. In a straight race they do not have the speed or stamina of many of the fleet-footed herbivores. The cheetah, on the other hand, does have the speed to run down its prey, such as Thomson's gazelle. Surprisingly, in view of the abundance of the herbivores, these large carnivores account for only 1 or 2 percent of the total savanna biomass, and they consume only about 15 percent of the herbivore biomass each year. This relatively low rate of exploitation is insufficient to control the herbivores, which are probably therefore limited by food or water. The predators appear to have a superabundance of food, and the fact that they do not exploit it fully suggests that their populations are not influenced by the quantity of prey available, as for example are the lynx populations in the boreal forest biome, but by some other feature—perhaps the ease of catching the prey. In addition to the big cats the herbivores are also food for packs of African wild dogs.

When these predators have finished with their kill the scavengers move in. Jackals and Spotted hyenas appear even

▲ **Searching for termites** and ants in a tree, a Southern tamandua (*Tamandua tetradactyla*). This anteater inhabits savanna habitats, among others, in northern South America. They may spend more than half their lives in savanna trees as they are clumsy movers on the ground.

▼► **Tidying up the savanna.** BELOW Dung beetles, such as these *Scarabaeus aeratus* females, feed on the droppings of other animals. They also collect balls of dung, as here, and place them in chambers in the ground, upon which they lay their eggs. RIGHT Marabou storks (*Leptoptilos crumeniferus*) feed on carrion and offal from animals that have died either naturally or through the actions of predators. They often roost, as here, in the savanna trees.

before the predators have finished feeding, and often drive them away. Both these species are, however, quite capable of catching their own prey. Other scavengers arrive, including the vultures and Marabou storks, and within two days there is little remaining evidence of the kill.

Savannas are rich in smaller species. There is a great range of bird species, many of them seed eaters like the quelea which, by diverting its attention from native grass seeds to introduced cereal crops, has become a considerable pest to farmers (p132).

Although the most apparent herbivores are the large mammals, insects are also important. The locusts have in the past caused immense damage to both natural and crop plants. They are indiscriminate feeders, eating almost any green vegetation in their path, and when they roost their sheer weight breaks branches. Locust swarms begin when prevailing winds bring individuals together to a site which has had recent rain. The females lay their eggs in the damp soil and the emerging hoppers crowd together on the succulent new plant growth. Crowded together like this the solitary grasshoppers change physiologically, morphologically and behaviorally into the swarming locusts. So different are these two phases that for a long time they were considered to be two separate species. Continued favorable weather conditions lead to a rapid buildup of numbers. Swarms the size of the Greater London area, about 1,000sqkm (400sqmi), have been recorded which, each day, consume enough food to feed the population of Greater London (approximately 12 million) for two weeks.

Termites are important elements in the decomposer chain, and their mounds are characteristic features of the savanna. They process large quantities of dead plant material, taking it below ground where it is either digested by microorganisms in the termites' guts or used by the termites as a substrate for fungi farming.

White settlers in Africa have looked to the savanna as a rich grazing land for their cattle. Nomadic tribesmen have herds of native cattle and sheep, but there are great difficulties in using high-yielding introduced stock. Disease, competition by native grazers, and poor grade fodder have forced a reappraisal of using the naturally-occurring herbivores. Game ranching using Thomson's gazelle and eland has been shown to be a profitable enterprise, as has controlled cropping of free-range animals. Whether such schemes become established will depend on demand and on strict legislation to prevent overexploitation. PDM/BDT

TEMPERATE GRASSLAND

Location and extent. . . Climate and vegetation. . . Productivity. . . Soil types. . . Large herbivores. . . Small mammals. . . Predators. . . Interference by man. . . Competition between herbivores. . .

LYING in the heart of the Northern Hemisphere continental masses, between the latitudes of 30° and 50°N, are the temperate grasslands—the steppes or prairies. In the Southern Hemisphere they are not so well represented, being found mainly in Argentina, where they are called pampas. Perhaps the tussock grassland of the South Island of New Zealand should also be included in this biome.

The climate in which the temperate grasslands have developed is typically dry, with hot summers and very cold winters. The further east one moves into Asia, the drier the steppes become, eventually blending into the aridity of the Gobi Desert. The western edge of the steppes may receive 400mm (16in) of precipitation, but in the east this can fall to 60mm (2.4in) or less. Similarly, the seasonal range of temperature becomes more pronounced in the east. The western Asiatic steppes have summer monthly average maxima of around 20°C (68°F) and winter monthly averages only just below 0°C (32°F). In the east, however, the summer temperatures attain monthly maxima of 25°C (77°F), and in winter the monthly minima can fall below −15°C (5°F). So the eastern steppes of Asia become a cold desert.

The east–west climatic gradient found within the Asiatic steppes is reflected in the vegetation. The western steppes are relatively rich in species. In spring there are wet areas within them where the melting snow accumulates, and these may even become covered by shrubs and small trees, particularly poplars and aspens. A similar "aspen parkland" is found along the northern edge of the North American prairies. Moving into the more arid steppes, however, the tree component is lost, as is the characteristically rich springtime flora, and the overall diversity is reduced.

The biomass of the temperate grasslands varies not only in space, but also in time. During the hot, dry summer it may become very small, but in its peak growth period, usually in spring, it can reach levels of about 3kg/sqm (0.8lb/sqft). This is only about half that of the tropical grasslands, and most of the temperate grassland dominants are far smaller in stature than their tropical counterparts; even the richest steppe grasslands have their upper canopy at only about 400mm (16in) above the soil surface. Their productivity is quite high, considering their low biomass, with average annual values around 0.5kg/sqm (0.12lb/sqft), and with some values as high as 1.5kg/sqm (0.37lb/sqft) on record.

Some studies in the North American prairies have shown that there is a distinct sequence of growth and productivity among the various grass species present. Those species which conduct normal photosynthesis and which are most successful under relatively cool, moist conditions grow mainly in the early part of the summer. Then, as the temperature rises and the water table in the soil falls, they are replaced by species more highly adapted for these stress conditions, particularly those grasses and other herbs which have evolved the modified photosynthetic system (called C_4 photosynthesis) which is so characteristic of the tropical grassland species.

The soils of the temperate grassland are very distinctive. Since the summers are warm and dry, the overall pattern of water movement in the soil profile is upward due to evaporation. Only during snow melt in the spring is this process reversed. The result is that soluble salts are transported upward through the soil, and often become concentrated in the upper horizons. Just how concentrated these compounds (which include lime, salt and gypsum) become, and how close they are to the surface organic layers of the soil, depends on the relative importance of the downward water movement in spring and the upward movement in summer.

One other feature of the soils of this biome, which has proved of great consequence to man, is their mobility. The vegetation is shallow rooted and the outcome of severe drought or overgrazing can be the loss of even this weak binding force. The winds which sweep this treeless biome may then pick up the loose surface soil and transport it through the air in violent dust storms. The redeposited soil is called *loess*, and the erosion

which results from such occurrences can have immense ecological and social effects upon agricultural lands and human populations within these areas, as in the "dust bowl" of the southern prairies of North America earlier this century.

Grasslands are preeminently the domain of grazers. Large grazing herbivores make great demands on the scanty vegetation at certain times of the year, and so tend to be migratory, moving from place to place to find new pasture. Small herbivores rely on stored food to provide for local adverse periods during the year.

The temperate grasslands share with the tundra the characteristic of being uniform, almost featureless and very exposed to the elements. In such landscapes, animals are often very obvious to predators, and one common feature of the fauna of these grasslands is the range of mechanisms adopted to reduce their apparency to other species.

Each of the three main geographic areas containing temperate grassland has its own selection of large herbivores that are ecological equivalents, reflecting the different evolutionary influences on the areas and their faunas. The Great Plains prairies of North America were in earlier times dominated by bison, and the plains of Europe and the Russian steppes by the European bison or wisent. These species rely on their large size and on herding for protection, and if roused can be aggressive. If alarmed, herds stampede, trampling everything in their path. They have poor vision but acute senses of smell and hearing. With so many noses and ears sensing the surroundings, the herd is not easily surprised. Big size brings other advantages; the surface area to body volume ratio decreases with increasing size, so that large animals have relatively less surface area to be chilled by the cold winter conditions. Indeed there is a general principle—though with many exceptions—called Bergmann's rule, that for any group of similar species those occupying colder climates will be bigger than those from

▼ **Herd of pronghorns** grazing in Wyoming, USA. The pronghorn (*Antilocapra americana*) inhabits the open prairies of North America, where it eats grasses, herbs and shrubs, migrating with the seasons to find the best forage. Like the American bison it became nearly extinct because of hunting, but now the population of pronghorns has increased to around half a million, at which level recreational hunting has little impact.

warmer areas. Big size also allows for a large stomach in which bacterial fermentation takes place. The bigger the capacity, the longer plant material can stay in the stomach being digested so that poorer types of forage are adequate to sustain the animal. Because of this, bison are unselective herbivores, producing an evenly grazed sward.

Unfortunately both species of bison have been hunted to the verge of extinction by man. In North America small numbers of bison have always been hunted by the Indians. With the introduction of the horse by the Spaniards, and later the use of firearms, the bison's defences were breached. The government of the time actively encouraged the killing of bison for trophies and their skins. For 30 years the slaughter continued: about 75 million animals were killed, and by 1880 there were almost none left. Several subspecies had been made extinct, but the few remaining individuals were afforded protection and the numbers today are increasing. Firearms and the growth of the human population in Europe had a similar effect on the wisent. The numbers of wisent dropped to six, but in a concerted effort to save the species they were brought together and encouraged to breed. Today about 2,000 wisent survive, all descended from that original nucleus of six animals.

The other large herbivore of the prairie is the pronghorn. In Europe and the steppes Red and Roe deer are common grazing species. Further eastward, in the drier, harsher regions of the steppes, are the remnant populations of the once widespread Saiga antelope and the only remaining race of the wild horse, Przewalski's horse.

On the South American pampas there are no really large herbivores and only one small deer, the Pampas deer, similar in size to the Roe deer. The only other sizable herbivore is the guanaco, a relative of the camels. As with the other temperate grasslands, the pampas has been modified by man's management and ranching exploits, but it seems that it has never supported the natural abundance of large grazers documented for the Northern Hemisphere grasslands. Instead there is a proliferation of large rodents like the Plains viscacha and the Patagonian cavy or mara, as well as the tuco-tuco, a small relative of the coypu. These rodents gain their protection by digging extensive burrows and tunnels. Their activities in stirring and mixing the different soil layers have an important influence on the vegetation, and serve to reduce the concentration of salts at the soil surface. During the summer many species of insect are active, particularly grasshoppers and ants. Several insectivores take advantage of this rich invertebrate life, including the Hairy armadillos.

Small mammals are also important elements in the Northern Hemisphere grassland communities. Black-tailed prairie dogs are abundant, living in large colonial burrow systems. They do not store food for use during the winter, but hibernate. Prairie dogs have a complex social life, with a language of gestures and greetings. There are several advantages to living communally: there are many watchers for signs of approaching danger; and the area around the colony is grazed down, preventing a predator from approaching too close without being seen. Individual colonies are separated, and this spacing may reduce overexploitation of the forage. Other burrowing small mammals of the prairie include the Black-tailed jackrabbit, the White-footed mouse and the Thirteen-lined ground squirrel.

The steppes of Asia also have a range of small burrowing mammals. The prairie dog equivalent is the European souslik, a close relative of the Thirteen-lined ground squirrel. The common hamster hoards vast quantities of seeds, nuts and roots which it uses during warmer periods in the winter. Two other important burrowers are the Steppe lemming, which feeds on vegetation, and the Common Russian mole-rat, which is a deep-burrowing root feeder.

All three temperate grassland areas have similar types of predators. The largest are wolves, coyotes and foxes, which feed principally on the smaller mammals and nesting birds. Wolves may cause some mortality among young deer in the spring. Smaller predators include weasels, stoats, polecats, badgers, and birds of prey.

The complex interactions of species that comprise the temperate grassland community have been increasingly interfered with as man exploits these areas for agriculture. Often the natural grassland community and agriculture do not mix. The burrowers damage crops, the larger grazers compete with domestic species, and predators threaten the survival of livestock. The killing of wolves and foxes leads to an increase in the numbers of small burrowing species. Their numbers also tend to increase if the productivity and diversity of plant life is improved by the growing of crops. Expensive extermination programs are then needed to control the burrowers. Without burrowers the soil structure begins to deteriorate. Overgrazing by domestic species and the practice of leaving the land devoid of vegetation between crops results in the loss of the thin topsoil during winter storms. Such destruction of the habitat has occurred in areas of both the Great Plains of North America and in the Central Asian steppes through a lack of understanding of the interactions that have evolved to maintain the temperate grassland biome. PDM/BDT

◄ **South American "ostrich."** Common or Gray rheas (*Rhea americana*) are large flightless birds of the South American grasslands. Although largely herbivorous, they do eat insects and other small animals.

► **Predator of the prairies.** The American badger or taxel (*Taxidea taxus*) will often dig out prairie dogs and ground squirrels from their burrows, and it is more carnivorous than the Eurasian badger. This one has caught a rattlesnake.

▼ **No place for nests.** The absence of trees on temperate grasslands limits the range of nesting sites for birds. Burrows are used by some species, such as these burrowing owls (*Athene cunicularia*), which take over those of rabbits and prairie dogs. Most other prairie birds have to nest among the grass.

Sharing the Spoils

How different sizes of herbivores live in a grassland

Why is only a certain number of species found in a given environment? This has been a question facing ecology since its origins. Explanations based on the availability of resources (such as food and light), competition between species for resources, fluctuations in climate, chance colonization by some species, and the numbers of predators, have been put forward singly or in combination. This question has been debated more perhaps for generalist herbivores (animals that feed on a wide range of food plants) than for any other type of organism.

The debate has centered upon whether the population numbers of these animals are limited chiefly by food availability, and whether the different herbivores inhabiting the same environment compete for food. This question is particularly relevant to the grassland environment, because grasslands contain the greatest diversity of herbivore species. To examine the evidence, a study was conducted on 16 herbivore species at the National Bison Range, Montana. These herbivores range in body size from 0.3g (0.01oz), a grasshopper, to 635kg (1,400lb), a bison. Fourteen other species occur within this range of body sizes (three additional common grasshoppers, meadow mice (three species), Ground squirrel, Cottontail rabbit, marmot, pronghorn, White-tail deer, Mule deer, Bighorn sheep and elk). Two data sets were collected: one dealing with the herbivore populations, and the other with each species' foraging behavior.

What, then, were the criteria for food limitation and competition? If the populations are larger when more food is available, the population size must be limited by the availability of food, and competition is indicated if the population is larger when the species is alone than when it is with other species. To find this out, population numbers must be manipulated either by removing animals from the environment or by establishing populations within enclosed areas.

Populations of large animals are hard to confine, and their removal is normally prohibited because of conservation considerations. Grasshoppers are much more suitable for use as model populations. They are not protected species, and they are small enough to enable populations to be established in

▲ **Competitive coexistence.** How herbivores of different body sizes exploit food plants in a way which permits their coexistence. The abundance, size, and chemical composition of plants vary spatially over a grassland. Herbivores of different body sizes use plants with different characteristics, with the result that some plants are used in common by animals of all body sizes, and other plants are used by animals of only one body size. These differences arise from the physiological constraints on each herbivore's diet choice; they permit the competitive coexistence of herbivores of different body sizes. The background photo shows American bison (*Bison bison*) grazing in North Dakota, USA.

cages. Experiments conducted with four species of grasshoppers demonstrate that populations within cages are larger when more food is available, and that the population of one species is larger when the other three species are absent or when the cages are set up in areas where larger herbivores are not present. Therefore, these herbivores appear to be food-limited and to compete for food.

If herbivores compete with each other, why doesn't one species that is more efficient at feeding eliminate the other species from the environment? This is the ecological concept of competitive exclusion: no two species can have identical resource use and coexist in the same environment, if that resource limits their population numbers. Each species' foraging behavior throws light on this.

Foraging behavior studies determine whether herbivores choose their diets as "optimal consumers"—that is, whether their food consumption maximizes their chances of reproduction and survival. This can be accomplished by the forager seeking one of two alternative "goals" in its diet choice. The first goal is to maximize daily nutritional intake, providing the animal with the greatest energy available for reproduction and survival. This assumes that the availability of nutrients limits reproduction and survival. The second goal is to achieve necessary nutritional requirements in the least time, so that more time is available for the care of young, mating, and the avoidance of predators. This assumes that the time available for activities other than feeding limits reproduction and survival.

The effectiveness of a specific food item to a herbivore is determined by three factors: firstly, how filling it is (each animal's total daily consumption is limited by its stomach size); secondly, how long it takes to eat a given amount of it; and thirdly, how fully it satisfies the animal's nutritional requirements. Taking these factors into account, the observed diets of the 16 herbivore species appear to maximize their nutritional intake, as expected if the populations are indeed food-limited.

Because of the above limitations on a herbivore's diet choice, each herbivore must feed on plants that are sufficiently quick and easy to eat and digest. The digestibility of a plant is related to its chemical composition, increasing with nutrients and the absence of fiber, and in general cropping rate is related to the abundance of the food items (leaves, twigs, whole plants, etc) and their size.

The size of the animal is also important, however. Small herbivores such as grasshoppers must consume easily digestible plants, but the food items do not have to be abundant or large. Large herbivores such as bison, on the other hand, do not require highly digestible plants, but the items must be both abundant and large. Intermediate-sized herbivores require intermediate digestibility, abundance and size. This body-size relationship emerges because small herbivores possess high nutritional demands per unit of body weight, and these requirements must be satisfied by a small digestive capacity, while large herbivores have relatively lower nutritional demands per unit of body weight (though still larger in absolute terms) that must be satisfied in a limited feeding time.

These differences lead to sufficient differences in resource use to permit different sizes of herbivores to live together, even though they compete. While different-sized herbivores may use and compete for many of the same plants, each may also be able to use a unique set of plants, which permits competitive coexistence. These findings explain why herbivores of different body sizes inhabit the region.

Similar explanations may account for the observed herbivore diversity in the East African savanna and other grasslands. Certainly these studies indicate the need to maintain a diverse plant resource if herbivores of different sizes are to persist in parks.

GEB

TUNDRA

Climatic conditions. . . Vegetation. . . Soil types. . . Productivity. . . Migration patterns. . . Insect life. . . Population cycles in lemmings. . . Winter survival. . . Ecological problems in a cold climate. . .

THE polar regions of planet earth experience high atmospheric pressure as the air is chilled and descends groundward. As a result, there is little precipitation in these areas and a cold, desert-like climate is produced. The very low levels of solar energy input, however, especially in the polar winter, mean that temperatures remain low. Evaporation rates stay correspondingly low, so water supply is unlikely to be a problem except on very well-drained soils in summer. Most of the tundra has wet, often waterlogged soils.

The tundra biome encircles the Arctic Ocean in the Northern Hemisphere, but consists only of scattered fragments in the Southern Hemisphere, mainly on the southern tip of South America and the various oceanic islands such as the Falklands. Its climate varies according to whether it is in an oceanic or a continental region. For example, the thin strip of tundra in the north of Europe, which lies mainly within Norway and the Soviet Union, is kept warm by the Gulf Stream drift currents from tropical waters. As a result, the Norwegian tundra does not begin until latitude 70°N, whereas in eastern Canada it commences around 55°N. These oceanic tundras may have several frost-free months during summer, whereas the continental tundra regions are never frost-free. Winter monthly minima may not fall below −10°C (14°F) in the west of Eurasia, but are often −30°C (−22°F) in Alaska and Canada, and in eastern Siberia monthly averages can reach −50°C (−58°F). The precipitation figures show a similar gradient, and usually fall within the range 150–300mm (6–12in).

The vegetation of the tundra regions consists almost entirely of perennial plants, the majority of which are hemicryptophytes and chamaephytes. There are some small shrubs, mainly willows and Dwarf birch, which are able to survive in the ice-blasted conditions of winter, when the sun does not shine and high winds loaded with ice crystals destroy any plant shoot which projects beyond the low vegetation cover close to the ground. The chamaephytes are more abundant here than in any other biome, because their cushion-like form is ideally adapted to these abrasive conditions. They include the saxifrages and various species of the heath and crowberry families (eg *Arctostaphylos*, *Vaccinium*, and *Empetrum*). Among the hemicryptophyte groups, the sedge family predominates, especially the sedges and cotton sedges.

The soils are often peaty as a result of waterlogging, but the alternate freezing and thawing of these peats can lead to large stones being forced up to the peat surface—the water freezes first below the (heat-conductive) stone and expands as it does so. Many of the arctic peatlands have massive ice lenses permanently embedded in the peat, which elevate the peat surface into mounds, often as much as 5m (16ft) in height and covered with a thick growth of lichens. All the soils are frozen below a certain depth, forming a "permafrost" layer which never thaws out. As the surface soil thaws it may become unstable on sloping ground and move downhill as a flowing sludge.

The cold and the low light intensities during the winter months have a pronounced effect upon the productivity of the biome. Its primary annual productivity is usually of the order of 0.1 to 0.4kg/sqm (0.03–0.1lb/sqft) which is comparable to that of semi-desert conditions. The biomass of up to 3kg/sqm (0.8lb/sqft) is also similar to that of desert scrub or of temperate grassland, but unlike the grassland the above-ground biomass is not replaced each year. It must be remembered, however, that although the productivity is small when expressed on an annual basis, most of this productive activity is confined to a very short growing season, often only three months. The tundra biome, therefore, is quite highly productive, but for only a very short period in each year. It is this which attracts migrant animals and birds to the high latitudes to breed during the short summer season.

The geological history of the tundra biome has been a checkered one. Before the succession of ice ages began, some two million years ago, the tundra was probably only just coming into being, as the global climate had been undergoing a prolonged cooling. Perhaps the mountainous parts of the earth provided some of the plants and animals suitable for colonizing the newly formed arctic desert. When each glacial expansion (and there have been many) was at its peak, the tundra conditions would spread into lower latitudes, only to contract again during the warmer interglacials. Sometimes the expansion of forest would almost completely obliterate the tundra habitat, but evidently the species involved survived and expanded their populations again with the next glacial advance.

The flat, almost featureless expanse of the tundra leaves animals exposed to the harsh elements. Most species avoid the

▲ **Barren landscape**—reindeer (*Rangifer tarandus*) grazing the tundra of Spitzbergen. Most reindeer migrate to the tundra to exploit the short flush of growth in the northern summer and return to the boreal forests in the winter, although these inhabitants are confined throughout the year to the island of Spitzbergen.

◄ **Impervious to winter cold,** an Arctic fox (*Alopex lagopus*) curls up and sleeps in its long white winter coat, which comprises 70 percent fine, warm underfur. The Arctic fox spends all its life in the tundra and has remarkable tolerance to cold. Its metabolism does not start to react to cold, for example by shivering, until the outside temperature is −50°C (−58°F).

worst tundra conditions in the winter. As the short tundra summer begins, there is an inrush of species from elsewhere. Many of the larger mammals of the boreal forests, such as the caribou, reindeer, Grizzly bear and Gray wolf extend their ranges northward during the summer. Reindeer in particular roam in great herds over the tundra in the summer, returning to the shelter of the forests during the winter. The birth of the reindeer calves is timed to occur in the early summer, just after the reindeer have moved out into the tundra. Following these herbivores, the wolves also give birth to their young in the tundra.

Virtually all the birds are migratory. Some, like the Willow grouse, move only short distances from the boreal forests, while at the other extreme the Arctic tern makes the incredible journey up from Antarctica. This bird, in making its twice-annual pole-to-pole journey, has the longest migration route of any species, and probably experiences the greatest amount of daylight and the least nighttime of any species on earth, since it visits both polar areas during the continuous days of their summers.

Of the hundred or so species of birds that visit the tundra during the summer to breed, the geese are perhaps the most dominant and characteristic. There are many species (eg Greylag, White-fronted, Snow, Canada, and Barnacle geese), but they all have similar lifestyles. They spend the winter in warm areas such as the Mediterranean, Africa or the southern USA and Mexico, and migrate back to the tundra to breed

during the summer, when there is sufficient food to raise young.

As the temperatures rise, the water in the surface soil melts but is unable to drain away because of the permafrost a short distance below. This boggy habitat provides the right conditions for many insects which have overwintered as resistant egg stages. Mosquito and blackfly larvae swarm in these surface waters, growing quickly on the rich particulate soup of peaty vegetation. Emerging as adults, later in the summer, they make life miserable for any warm-blooded animal. Female mosquitoes and blackflies need blood meals to provide the protein for egg production, and so they voraciously attack anything which might provide the blood they require. To escape these dense swarms of intensely irritating mosquitoes, caribou and reindeer move to higher, drier ground away from the low-lying boggy habitats where the blood-sucking flies congregate in greatest abundance. The mosquitoes make these tundra regions equally uninhabitable for humans.

While the mosquitoes plague the larger mammals, they are the very reason that many insectivorous birds migrate into the tundra. They breed there and rear their young on the abundant insect life.

This summer increase in the tundra population naturally attracts predatory species as well. Some, like the merlins and falcons, are attracted specifically by the many migrant birds, but many others move in to feed on the lemmings. The wolves come up from the boreal forests to have their young in the tundra because of these small mammals rather than because of the caribou. The comparatively rich growth of vegetation in the summer is a time of rapid growth and reproduction in lemmings. They may become sexually mature in as little as 15 days from birth, and give birth themselves to their first litter when only 38 days old.

▲ **Huddled together for warmth.** The only large mammal to stay in the tundra all year round is the Musk ox (*Ovibos moschatus*). It is protected from the bitter cold by long matted windproof hair over thick fat layers. These animals huddle together in steaming groups for warmth and for protection from predators.

► **Summer visitor,** the Arctic tern (*Sterna paradisaea*) nests in the tundra, but spends the rest of the year circling the Antarctic.

Lemmings show the characteristic four-year cycles of other small rodents in boreal areas (see the section on the boreal forest for a more detailed description) and these will naturally govern to some degree the reproductive success of the predators from year to year. During their years of peak population density, the lemmings may eat out all the available vegetation that forms their staple diet. At such times lemmings migrate to new areas. These migrations are spectacular and are all the more apparent because the lemmings may move into towns and villages and become a nuisance to the inhabitants. They tend to move towards lower ground. Lemmings do not migrate in a flowing river of bodies, but tend to be spaced out. Individuals

are aggressive to each other, with large females dominant in such conflicts. The majority of the migrant lemmings are young, small and immature.

Stories abound of lemmings throwing themselves to their death over cliffs into the sea during these migrations. Many of these stories are associated with mountainous terrain which is deeply divided by fjords. In such situations routes to other areas are limited, so that lemmings would be forced to move in only one or two directions, and migrations would therefore be far more obvious. Lemmings are good swimmers, and in fjord country swimming may be the only way forward. So it is also likely that they have been seen jumping into the sea or a river, not commiting mass suicide, but simply to swim to better feeding grounds.

As temperatures fall at the end of the summer, the birds leave for warmer habitats and many of the larger mammals move south to the shelter of the coniferous forests. Small mammals such as lemmings and Arctic hare stay, and so do some predators such as the ermine or stoat and the Gyr falcon. About the only large mammal to stay in the tundra all year around is the musk ox.

Hibernation is not possible in the tundra. The soil is frozen so that tunnelling is impossible and the summer season is too short to lay down sufficient food reserves to last the long winter. In depressions the snows collect and insulate the ground from the extreme cold. Small mammals tunnel through the snow to get food and may also breed during the winter. The stoat also follows the tunnels for its prey. Larger predators such as the Arctic fox lay up frozen food caches of dead animals to help them through the winter. Arctic hares shelter in snow burrows. They feed on the surface and risk being preyed on by the Arctic foxes. Several of these species—stoat, Arctic fox, Arctic hare and Willow grouse—change color to make themselves less obvious in the featureless tundra. Their summer coats are in shades of brown, while the winter coats are pure white.

Coping with the rigors of the climate is all-important in the tundra. As the higher latitudes are approached the battle for survival subtly shifts in emphasis. In the humid tropical rain forests, species have evolved complex mechanisms for camouflage and defense to protect them against other species. The forces acting to mold the lifestyles of species in this biome are all biological. Toward the poles the biological forces become less important, and the pressures of climate become the dominant influences. Only a relatively few species can survive in the harshness of the tundra—and that goes for predators too.

Some ecologists have commented on the fragility of the simple foodwebs of tundra communities. However, modern ecological opinion is changing to the view that these simple communities are in fact adapted to be able to withstand enormous disturbances or perturbations, since these are part and parcel of their natural experience. The tundra biome is constantly subject to violent changes in climate, and these are often of an unpredictable nature. Whether tundra communities are more or less stable than the tropical rain forest communities is still a hotly debated topic among ecologists, with one of the major problems being the definition and measurement of stability.

PDM/BDT

MOUNTAINS

The effect of altitude on climate and flora. . . Mammal species. . . Insect life. . .

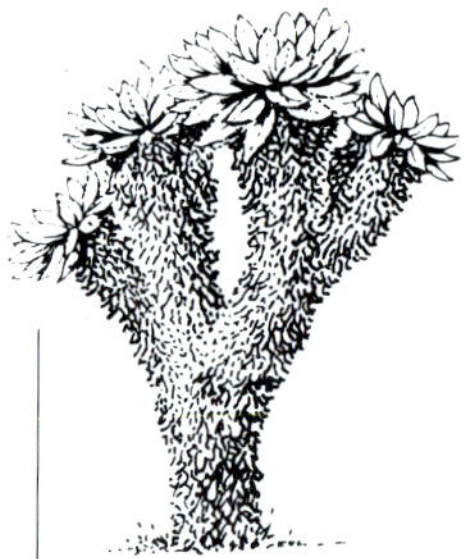

MOUNTAINS—and particularly the highest mountain ranges—are among the most spectacular of geological features. Mountainous regions are areas scattered around the world which have been affected by geological events, particularly the collision of continental plates. This creates a buckling of the surface crust, producing elevated ridges of mountainous land. Such raised areas can so affect the local climate that the usual biome of the particular region is replaced by others normally associated with cooler latitudes. Sometimes a high mountain in the equatorial region may bear a sequence of altitudinal zones which resemble (at least in their life-forms) the biome sequence as one proceeds polewards.

The Himalayan mountain range, for example, has its southern feet within the savanna-clothed plain of northern India, but its foothills, which collect additional waters from the summer monsoon, are covered with a subtropical deciduous forest of sal and arjun, or terminalia tree, together with some teak. Ascending the foothills, this gives way to a more temperate type of deciduous forest of oak and rhododendron above about 1,400m (4,600ft). Around 2,500m (8,200ft) this grades into forest with an increasing coniferous component, notably

▼ **Gliding over the Himalayas,** a lammergeier (*Gypaetus barbatus*) surveys the terrain for food. Like several other mountain-inhabiting birds, the lammergeier, or Bearded vulture, is a scavenger feeding on carrion. It is known to break open bones by dropping them from great heights, cracking them so that it can get at the marrow.

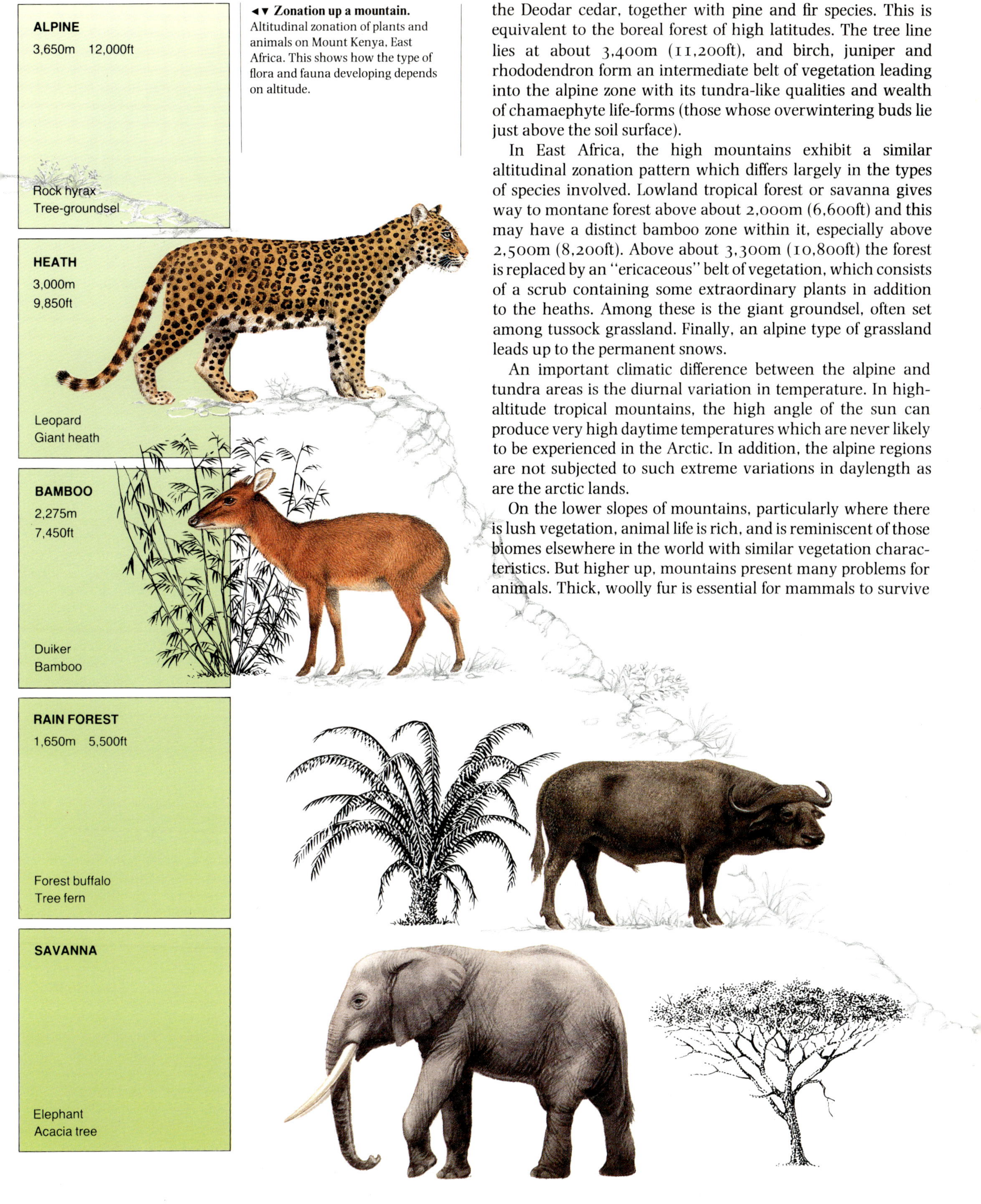

◄▼ **Zonation up a mountain.** Altitudinal zonation of plants and animals on Mount Kenya, East Africa. This shows how the type of flora and fauna developing depends on altitude.

the Deodar cedar, together with pine and fir species. This is equivalent to the boreal forest of high latitudes. The tree line lies at about 3,400m (11,200ft), and birch, juniper and rhododendron form an intermediate belt of vegetation leading into the alpine zone with its tundra-like qualities and wealth of chamaephyte life-forms (those whose overwintering buds lie just above the soil surface).

In East Africa, the high mountains exhibit a similar altitudinal zonation pattern which differs largely in the types of species involved. Lowland tropical forest or savanna gives way to montane forest above about 2,000m (6,600ft) and this may have a distinct bamboo zone within it, especially above 2,500m (8,200ft). Above about 3,300m (10,800ft) the forest is replaced by an "ericaceous" belt of vegetation, which consists of a scrub containing some extraordinary plants in addition to the heaths. Among these is the giant groundsel, often set among tussock grassland. Finally, an alpine type of grassland leads up to the permanent snows.

An important climatic difference between the alpine and tundra areas is the diurnal variation in temperature. In high-altitude tropical mountains, the high angle of the sun can produce very high daytime temperatures which are never likely to be experienced in the Arctic. In addition, the alpine regions are not subjected to such extreme variations in daylength as are the arctic lands.

On the lower slopes of mountains, particularly where there is lush vegetation, animal life is rich, and is reminiscent of those biomes elsewhere in the world with similar vegetation characteristics. But higher up, mountains present many problems for animals. Thick, woolly fur is essential for mammals to survive

► **Hanging to the mountain edge**—a group of Dall's or Thinhorn rams (*Ovis dalli*) resting, with Mount McKinley, Alaska, in the background.

the penetrating cold, and some of the most highly prized furs and wools come from mountain-dwelling animals like the chinchilla and alpaca from the Andes of South America, and Angora goats from the mountains of southern Central Asia, from which we get mohair. Larger mammals have to be very sure-footed in the rocky terrain. Some, like the chamois and the Siberian ibex, are legendary for their ability to move rapidly over terrain that appears impassable. Larger still, the Tibetan yak is restricted to less precipitous areas. This species is semi-domesticated and is now rare in its natural habitat.

Large predators are scarce, and lead a difficult life in the mountains. The rare and beautiful Snow leopard lives permanently above the snowline in the highlands of eastern China and the Himalayas, preying on ibex and tahr. A study of Gray wolves on Mount McKinley in Alaska illustrated the essentially scavenging role of these carnivores. The wolves feed on Dall's sheep, and the sheep skulls left on the mountainside can be aged by cutting through the horns and counting the growth rings, in much the same way as trees can be aged. A large number of skulls were examined, and they showed that only very young or very old sheep were killed by wolves. Dall's sheep in their prime can easily escape from a wolf.

At high elevations wind is a considerable problem, adding to the chilling effect of the cold; it restricts flight in all but the strongest of fliers. One such is the condor which soars on the wind currents over the entire length of the Andes. The lammergeier is the Eurasian equivalent: it feeds on bones which it breaks up by dropping them from some height onto the rocks. Smaller birds keep close to the ground and often burrow to avoid the worst of the cold at night. The Andean Diuca finch gathers in tightly-packed flocks, and shelters under rocks to keep warm.

Surprisingly, insects may be quite abundant at these high altitudes. The winds blow many lowland insects up into the air, freezing them and depositing them on the mountain snow fields. Here they are preyed on by the resident insects and spiders, some of which are very unusual. For example, the grylloblattids are restricted to the snow fields of the Rocky Mountains of North America, and are found nowhere else. They are wingless and look like a cross between a cricket and a cockroach; they may be the isolated remnant of an order from which the present-day grasshoppers and cockroaches evolved. They are active at temperatures around freezing point, and are killed by warmth. They scavenge over the snow fields, feeding on the frozen stranded lowland insects. Other insects that live permanently in this area are also often wingless, since directed flight for such small animals is not possible in the high winds. Huge assemblages of ladybugs often collect under rocks in the fall. They overwinter in the mountains, protected beneath a blanket of snow. In the summer, just below the snowline, the flowers of the alpine pastures are rich with butterflies.

The very inhospitality of most mountain areas is probably why many of its unique and interesting animals still survive. Of all the world's biomes, high mountain habitats are least affected by man and his activities. PDM/BDT

SALINE WETLANDS

Mangroves. . . Temperate swamps and saltmarshes. . . Animal life in saltmarshes. . . Invertebrate life. . . Mangrove fauna. . .

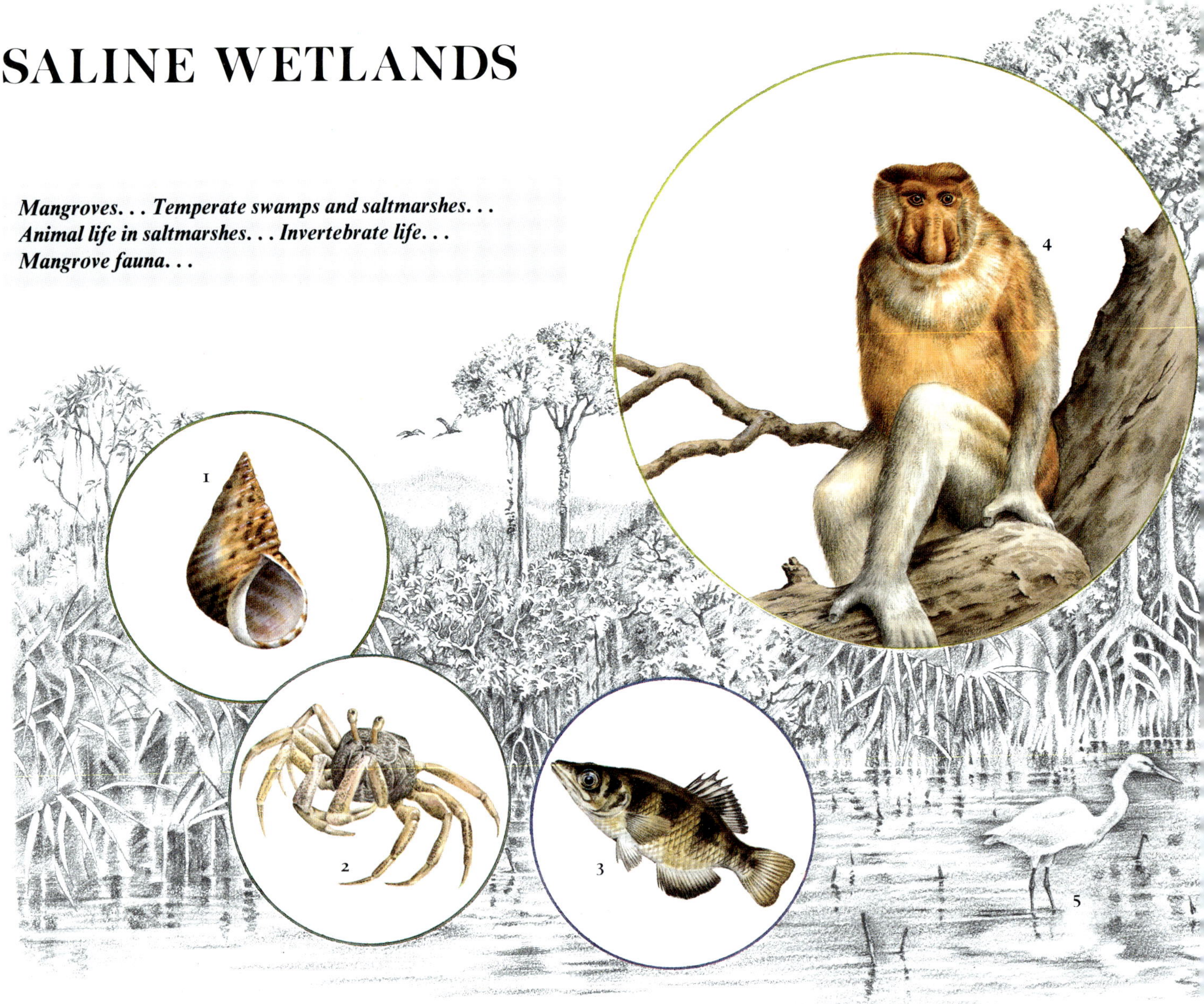

THE coastal zones of the continental land masses are often covered by a characteristic maritime vegetation, and where there is much sediment deposition from the flowing water, as in the estuaries of rivers, the development of saltmarshes, lagoons and swamps can result. These are the saline wetlands.

In the tropics, the most frequent type of saline swamp is dominated by the mangroves. There are many different kinds of mangrove, and they are adapted to a whole range of different conditions, such as variations in salinity or water depth, and to sandy or muddy sediments. All of them have a remarkable tolerance of high water tables around their roots, which is fatal for most plants because they are unable to respire properly and hence drown. The mangroves possess specialized breathing mechanisms in their roots, called *pneumatophores*, which enable them to survive.

Often the different types of mangrove form a series of bands or zones within the swamp, parallel to the shore. These seem to relate to the amount of salt concentration which the different mangrove species can tolerate. Genera such as *Sonneratia* occupy the outer zones because they cannot tolerate salt concentrations above that of seawater. The inner zones are occupied by more tolerant genera, such as *Avicennia*, for in these more protected waters evaporation can result in very saline conditions.

In the temperate regions the mangroves are replaced by grass-dominated swamps and marshes. In the Mediterranean area, for example, the estuaries of the Danube, the Rhône and the Guadalquivir form extensive deltas in which grasses like the reed and the Sea bulrush play a dominant role. Strongly tidal sites may bear a vegetation of chord grass or extensive flat marshes in which a number of low-growing perennial herbs form a uniform carpet, dissected by drainage creeks and with scattered saline pools. Many of these plants have specialized glands with which they excrete the excess salt that they have taken up from their environment.

The saline wetland is a habitat which is very well recorded in geological history, simply because the plant remains have survived well as fossils. Many of the Carboniferous swamps may have been of this type, and 60 million years ago the area around present-day London bore a vegetation not unlike some of the estuarine swamps now found in Southeast Asia.

Saline wetlands are principally habitats for small, non-specialist, often highly mobile species of animal. The yielding nature of the substrate in saltmarshes, and the densely-packed roots and trunks in tropical mangrove swamps, preclude most large mammalian species. There has been a history in Europe of using the upper portions of saltmarshes as grazing land for cattle and sheep, although this practice is not found to any degree elsewhere. The tides govern grazing by cattle and sheep as well as by smaller species such as the European rabbit over

most of the saltmarsh expanse. Species move out to gather food at low tide, and return to the higher levels as the tide returns. These higher levels receive considerable disturbance due to more intensive grazing, trampling and nutrient enrichment through the deposition of feces.

Birds are a dominant feature of saltmarsh ecology. A number of species, such as the redshank, oystercatchers, skylarks, seagulls and terns use European saltmarshes as breeding grounds. Others, particularly geese, use them as feeding grounds during their migrations. Brent geese feed on eel grass and saltmarsh grass while the Greylag goose and the Bean goose consume the roots of Sea club rush. Similar patterns of use are seen in the North American saltmarshes. In large numbers these herbivores may also change the composition of saltmarsh vegetation.

Invertebrates are common in saltmarshes, and form an important food resource for birds, especially in the summer when insects reach their greatest abundance. Since very few spiders and insects have been able to colonize seawater, they are restricted to the area above the tidal zone, leaving the tidal zone itself populated by a variety of crustaceans. There is a small number of insects that have carved a marine niche for themselves. The aphid *Pemphigus trehernii* feeds on the sap taken from the roots of the Sea aster which grows in mud regularly inundated by the sea. Beetles of the genus *Bledius* also

◄▲ **Animals that occur in mangrove swamps** in different regions of the world. (1) Mangrove winkle (*Littorina scabra*), Seychelles. (2) Soldier crab (*Dotilla mictyroides*), SE Asia. (3) Archer fish (*Toxotes jaculator*), SE Asia. (4) Proboscis monkey (*Nasalis larvatus*), Borneo. (5) Great egret (*Egretta alba*), worldwide. (6) American darter (*Anhinga anhinga*), N and S America. (7) Saltwater crocodile (*Crocodylus porosus*), SE Asia, N Australia. (8) Scarlet ibis (*Eudocimus ruber*), S America, W Indies. (9) Mudskipper (*Periophthalmus* species), SE Asia.

occupy the areas of the saltmarsh exposed between the tides.

The mangrove swamps of the tropics are poorly understood. These areas are used by many marine species as spawning grounds, and are rich in invertebrate life. Crabs, including the fiddler crabs, are common, scavenging over the exposed mud. They are frequently territorial, and the fiddler crab uses its oversized claw to gesticulate at its neighbors. Mud skippers, fish which emerge from the water to feed on the mud surface, also set up temporary territories on the mud during the mating season. In more open areas crocodiles bask on the mudflats.

As in the saltmarshes, mangroves support large numbers of birds. Most of these species feed on the abundant fish and invertebrate life, but the bizarre hoatzin, which inhabits the mangroves of the Amazon, feeds on leaves of mangrove and arum.

PDM/BDT

FRESHWATER WETLANDS

Vegetation and productivity... The development of fens... Large herbivores... Small mammals... Birds... Reptiles and amphibians... Adaptation to climate...

WHEREVER bodies of freshwater accumulate on the face of the earth, a process of sedimentation begins which tends, in time, to fill them up. If wave action is not too fierce, then aquatic plants begin to grow around their edges where the water is shallower. These may be of an upright stance, called *emergent* species, or they may be floating or submerged *aquatic* plants. If they root in the basal sediments, then they are not likely to be able to grow in water of any great depth; about 5m (16ft) forms a limit for many. But free-floating species are not limited in this way, and some, like the water fern and the water hyacinth, form extensive floating carpets on tropical lakes—they may even block waterways.

In many African lakes, a fringing swamp of papyrus forms floating mats, and parts of this can break off and float out into the lake as islands. These papyrus swamps are very productive, with annual primary productivity values of about 2kg/sqm (0.5lb/sqft) and with some observations as high as 10kg/sqm (2.5lb/sqft). It must be regarded, therefore, as one of the most productive ecosystems in the world.

In the Amazon flood plain in Brazil, swamp vegetation is extremely well developed. There is a high precipitation of about 2,000mm (80in) per annum, and temperatures are consistently high at around 20–22°C (68–72°F). So conditions are ideal for growth, and productivity is high. Over the huge Amazon catchment there is some seasonal variation in the amount of precipitation, so that there is a fluctuation in the level of the water. In the Rio Negro, a tributary of the Amazon, the water level may fall a total of 10m (33ft) or more between June and December, so the swamp vegetation has to adapt to these very large fluctuations. The floating aquatics are perhaps ideally suited to this, since they simply follow the water level. Among them is the impressive Royal water lily of the Amazon with leaves up to 1.8m (6ft) across. Like the African papyrus swamps, there are also floating mats and islands of vegetation, especially of grasses such as *Pasalpum* and *Echinochloa*, and these too have high annual productivity, with 3kg/sqm (0.8lb/sqft) recorded. Similar results have been obtained from the Saw grass swamps of the Florida Everglades.

In the temperate regions, swamps are often dominated by a single emergent species, such as the Reed mace or the Saw grass, and this has the effect of reducing the overall diversity of the plant association. But even in temperate swamps the annual productivity is high, even as high as 2.5kg/sqm (0.6lb/sqft), which compares well with the best that can be achieved by temperate deciduous forest.

The growth of emergent aquatic plants influences the pattern and speed of water flow through the swamp,and leads to a more rapid accretion of the silt and sediment around the shoot bases. This, in time, has the effect of raising the sediment level, and the water becomes shallower until eventually it may be above the water table during dry spells in summer. The swamp has now turned into a fen, and this is often accompanied by a great increase in the diversity of the plant species present. It may also result in the invasion of tree species, especially willows and alder. Once the trees have established a fen woodland, the structural complexity of the habitat is greatly increased, which offers new opportunities for animals, especially invertebrates. Estimates of the primary annual productivity of fen woodlands are of the order of 0.5 to 0.8kg/sqm (0.1–0.2lb/sqft) in boreal sites and up to 1kg/sqm (0.25lb/sqft) in temperate areas, but the productivity tends to be less than that of the swamps and fens which precede them in the successional sequence.

In northern temperate areas, the succession may proceed further by the invasion of sphagnum mosses into the fen woodland, which leads to the elimination of the trees and the takeover of bog. The sphagnum mosses absorb all the nutrients from the water and make their environment very acid, so the tree seedlings can no longer establish themselves. The productivity of the sphagnum bog is even lower than that of the fen woodland, with annual values around 0.1 to 0.3kg/sqm (0.025–0.075lb/sqft). Treeless bogs of this type are mainly confined to very oceanic areas, such as the extreme west of Britain; in continental regions they are often colonized by coniferous trees and can be regarded as part of the boreal forest biome.

Despite the great variety of freshwater wetlands in the world they all present similar problems for the animals living in them. These problems are mainly associated with seasonal changes in the water level. Too high a water level and the more terrestrial species are deprived of their habitat: too little water and the marsh dries up, creating desiccation problems for the species adapted to moist or wet environments.

By their nature, marshes and bogs are unsuitable habitats for most large mammals; but in the African papyrus swamps the hippopotamus is an obvious exception. Groups of these large animals, which may weigh as much as 2.5 tonnes each, spend the day wallowing in the muddy water. They stir up the mud and nutrify the water with their feces, to the considerable benefit of the plant and aquatic life. In the water the hippopotami feed on Water cabbage, and during the night they come out onto dry land to graze the surrounding grassland. This nocturnal grazing is considerable. One study estimated that each hippo takes about 140kg (310lb) of herbage nightly. Hippos cut the grass very close to the ground, so that overgrazing may occur to the detriment of other species. Buffalo numbers increased in the Queen Elizabeth National Park in Uganda when the hippo numbers were reduced by cropping.

In these African swamps there are several smaller semiaquatic grazing mammals such as the sitatunga and the lechwe whose hooves splay out widely to prevent the animals sinking into the mud. In South American swamps the largest of the rodents, the capybara, is one of the principal herbivores. It swims well, spending much of its time in the water. As in the hippo, this aquatic habit has led to the ears, eyes and nostrils being placed along the top of the head.

North American swamps are populated by far smaller mammals, the most characteristic being the muskrat, a large water vole also commonly associated with ponds and lakes. It is omnivorous, feeding on vegetation, snails, crustaceans and the occasional fish. These rodents make large lodges of vegetation in which they live in small family groups. There may be two to three lodges per hectare (0.8–1.2 per acre), each containing

▲ **Alligators in the Okefenokee Swamp,** Georgia, USA—American alligators (*Alligator mississippiensis*) basking on a mud bank.

about five animals. At greater densities the muskrats overgraze the marsh vegetation. Another vegetarian, the Swamp rabbit, does not jump (as do its terrestrial relations); instead it is well adapted to the marsh habitat, being able to swim well and move through soft mud.

The major mammalian predator of these herbivores, especially of the muskrat, is the American mink, an aquatic member of the weasel family. It is an excellent swimmer and will also feed on fish, frogs, snakes and waterbirds. Owing to the high quality of its fur, the mink has been imported and farmed in other areas of the world, including Great Britain. Escaped mink from these farms have established themselves in a variety of habitats.

When the water level drops, several other common small mammals, including the Common raccoon, invade the marshes. Red foxes and coyotes are also attracted to the rich pickings in the marshes during the dry season.

In Europe the wetlands are predominately bird habitats, the marshy ground forming an effective barrier protecting the birds from most terrestrial predators. Mammals are not very common—although, as in North America, several species of carnivores move into the marshes during the summer when the water level drops and the ground dries out. The nesting birds and their offspring are very vulnerable during such times, and many may be killed. In one of the most extensive and impressive wetlands of Europe, the Coto Doñana of south western Spain, the breeding birds show site differences in where they nest. Some, like the European avocet, prefer the raised islands that are dotted throughout the marshes. Others, like the large Marsh harrier or the coot, nest in the sedges and bulrushes, while Purple herons go for the reeds. All the major wetlands of the world support a variety of large birds, indicating the richness of these habitats as feeding grounds. Each species, however, has a distinct feeding niche which can be readily appreciated by a consideration of the differences between species in bill size and structure and the length of leg. Long legs enable birds to exploit the fauna living in deep water, while the short-legged species are restricted to the shallows. Long, pointed beaks are used to stab at fish and frogs or to probe into the mud for worms and mollusks; the "inverted" beak of flamingos sieves small crustaceans; and the curved beak of the avocet and the spoonbill's spoon-shaped bill are swept through the water or soft mud to collect invertebrates.

Marsh lands are rich in reptiles and amphibians, particularly in the warmer tropical and subtropical areas. The largest reptiles, the alligators and crocodiles, are important top carnivores feeding on both aquatic and terrestrial prey. Snakes are one of the few terrestrial predators to be able to move relatively easily through the swamp vegetation, and several species are adept swimmers. Their main prey consists of lizards, amphibians, insects, nestlings and, in the drier areas, small mammals. As with the birds, the snakes hunt in different ways and in different parts of the habitat, and so the relative proportions of the prey items varies from species to species. In the Coto Doñana there is a complex foodweb within the rich reptile fauna, with the invertebrate-feeding lizards being preyed on by several species of snake, some of which in turn are also eaten by the large, venomous Montpellier snake.

The plants and animals of the wetlands are adapted to a life with plentiful water; however, unlike a lake or river, the wetlands frequently dry out in the summer. Wetland species must also therefore cope with lack of water. For the truly terrestrial species this presents little difficulty. Herbivores may have to move to find new grazing. The hippo may move up to 7km (4mi) away from the marshes to find its nighttime pastures.

But for the aquatic and semiaquatic species, desiccation is a major problem, because they lack the waterproof skin of terrestrial animals and lose water quickly. As the water level drops, the water becomes warmer and stagnant. The vegetation collapses down into a mat that reduces water loss from

▼▶ **Birds of Coto Doñana wetlands:** (1) Marbled teal (*Marmaronetta angustirostris*). (2) Ruddy shelduck (*Tadorna ferruginea*), male. (3) Marsh harrier (*Circus aeruginosus*). (4) Whiskered tern (*Chlidonias hybrida*). (5) Greater flamingo (*Phoenicopterus ruber*). (6) Cattle egret (*Bubulcus ibis*). (7) Purple swamp hen (*Porphyrio porphyrio*). (8) Black-crowned night heron (*Nycticorax nycticorax*). (9) Great reed warbler (*Acrocephalus arundinaceus*).

3

1

2

the mud. Fish relying on dissolved oxygen must either move to deeper, more open water, or die. One remarkable group, the lungfish, have developed a gill cavity which they fill with air after the manner of a "lung."

As conditions get worse, many small aquatic forms (mollusks, worms, insects and protozoa) encyst in the mud. The lungfish also burrow into the mud and surround themselves by a slime and mud cocoon, leaving a porous plug at the burrow entrance for oxygen exchange. As the mud dries out, the lungfish remain protected by the hardened cocoon. When the water level rises again there is a rapid recolonization as the quiescent forms break out of their protective mud-bound cysts.

In the cold northern temperate bogs, animal life is far less diverse. Three factors contribute to this paucity: the low temperatures, and the acidity and low oxygen content of the water. Insects, especially dipterans, and other invertebrates are common, and form an important food resource for migrant birds, but the high acidity rules out most mollusks, which need alkaline chalky conditions for shell production. The conditions are frequently too cold for reptiles and amphibians, and the only mammals tend to be the widespread muskrat and beaver. Perhaps the most typical bird is the crane. PDM/BDT

FRESHWATER

The variety of freshwater habitats. . . Ponds and lakes: their origins and chemical make-up. . . Flora and fauna of lakes. . . Rivers and streams: the effects of variation in water flow. . . River fauna and flora. . .

THE freshwater aquatic habitats include an enormous range of types, some immense, some tiny; some short-lived, some whose duration can be measured in geological time. A basic subdivision is into static bodies of water such as lakes, and moving water including streams and rivers. The variety of habitats and the consequent differences in opportunity for animals, can be illustrated by the fact that a pool in a fork of a jungle tree, a puddle left after seasonal rains in a semidesert, a garden pond, and the Great Lakes of North America are all examples of static water. A tiny field ditch, a rushing mountain torrent, and the Amazon are likewise all examples of flowing water.

Evidently such diverse habitats will present different problems and will contain many kinds of plants and animals with differing lifestyles.

Ponds and lakes are formed by precipitation filling a cavity. The cavity may have been created by major earth movements such as mountain building, by glacial or volcanic activity, by landslips due to local geological instability, or even by animal activity (eg beavers and man). Thus lakes may be of great antiquity (Lake Baikal dates back 25 million years) or of very recent origin.

The quality of the water filling the ponds and lakes depends on the local soils and rocks. If precipitation exceeds evaporation a stream or river may form at the lake's lowest point around the margin. As well as water, all ponds and lakes receive some material washed down from the surrounding land, some of which is deposited on the bottom while some remains in solution in the water.

Where this addition of solid material (sediment) is rapid, open water relatively quickly gives way to swamp, as plants colonize the sediment, alter the conditions by their own presence, and begin the progress along the so-called "hydroseral succession" from open water to damp forest. Most lakes and ponds are fringed with vegetation from some stage in the hydroseral succession; which stage depends on the age of the lake or pond, the rate of runoff, and so on.

Colonization of any water mass begins almost as soon as it is formed. Flying insects and microscopic creatures with small, highly-resistant eggs or cysts which can easily be carried by the wind usually arrive first, and colonization then proceeds gradually. As more possible habitats (chiefly weeds and mud) become available, so the variety of lifeforms increases until the system is ecologically saturated, and a community has been formed. The animal species present change as time goes on, the early colonists being replaced by other species as a result of competition, selective predation or changing conditions in the pond. Many small ponds dry up before a complex fauna can develop, but exist long enough to allow short-lived species to complete their life history, and some species may be found only in such temporary ponds.

The diversity and abundance of organisms in a lake will also be affected by the chemical quality of the water. Several technical terms based upon chemical differences are used to describe lakes. Thus there are *eutrophic*, *oligotrophic* and *dystrophic* lakes. These differ in the dissolved inorganic nutrients they contain, and hence in their productivity and some other dependent characteristics.

Eutrophic lakes are productive lakes, usually broad and shallow, whose input comes from base-rich soils. They have a high content of dissolved ions such as calcium, and will usually also contain quantities of nitrogen and other elements essential for plant growth. The organic content is high. Because of the large amount of plant growth, especially planktonic algae, in such lakes, the water is yellow or green in color and opaque. Near the surface, oxygen levels are high, particularly during photosynthesis, but deeper down the presence of much decomposing matter leads to oxygen shortage. This is characteristic of eutrophic lakes.

These water characteristics are very favorable to plant growth, and eutrophic lakes are usually fringed with lush vegetation. Calcium is important to some types of invertebrate (mollusks, worms, flatworms and crustacea are all more abundant in calcium-rich water) and these will be abundant in eutrophic waters. On the whole, however, the numbers of species of zooplankton and other fauna will be less in eutrophic waters than elsewhere.

Oligotrophic waters, by contrast, are derived from hard,

► **Nocturnal hunter.** The Freshwater crayfish (*Astacus pallipes*) hides under stones or in burrows in river banks during the day, and at night crawls over the bottom where it feeds on snails, worms and insect larvae.

▼ **Aquatic stroll**—a bull hippopotamus (*Hippopotamus amphibius*) on an African riverbed and surrounded by fish. Although hippos feed almost exclusively on land they defecate in lakes and rivers, thereby increasing the nutrient content of the water which promotes fish production. Some fish, such as the *Labio velifer*, seen here congregating around its snout, graze on algae on hippo skin.

base-poor rocks and contain only low levels of nutrients. They are relatively unproductive, often deep, very transparent, as there is little plankton or other suspended matter, and the bottom consists of mineral sediments, unlike eutrophic lakes where the productivity results in an organic substrate. Oxygen is abundant in such lakes at all levels. While production is low, and therefore the numbers of animals and plants are also low, species richness is often high. Groups of animals requiring calcium are infrequent or absent from such waters.

Dystrophic lakes are characteristic of acid soils such as heathlands or peat moors and are most obviously distinguished from other lakes by the color. They have brown water due to the presence of humic acids and undecomposed plant material. Calcium and many other ions are lacking, and conditions for animal life are poor. Few species (mostly insects) are found in such pools, and even then in small numbers.

In addition to these widely represented types, there are other pools with more extreme conditions.

In every pond or lake the fauna will consist of *plankton*, *nekton* and *benthos*. The plankton consists principally of small animals of 0.5–2mm (0.02–0.08in) in size: protozoa, rotifers, crustacea, water fleas and copepods for example, which graze the phytoplankton and are themselves a major food resource for predatory species. Many will have life cycles evolved to allow maximum usage of food resources; thus water fleas produce their young parthenogenetically and frequently in times of food abundance, and only reproduce sexually when conditions decline. Plankton may enter a resting phase during winter in temperate areas to escape from their predators and because the food supply is inadequate.

Such animals are fed on by fish—which constitute a large proportion of the nekton—and by some invertebrates. Many fish also feed on the animals on and in the bottom and in and around the plants. These benthic animals are from many phyla, including protozoa, flatworms, mollusks, crustacea and fish, but insects predominate, particularly in nutrient-deficient waters.

Benthic animals have a variety of food available to them. As well as the plants and animals of the plankton, there is the

▲► **Animals in and around a European lake.** A In shrubs: (1) Blackcap (*Sylvia atricapilla*). B In reed swamp: (2) Sedge warbler (*Acrocephalus schoenobaenus*); (3) European water vole (*Arvicola terrestris*). C In water where plants have floating or emergent leaves: (4) Common frog (*Rana temporaria*); (5) Gray moorhen (*Gallinula chloropus*); (6) Mayfly (*Cloeon dipterum*) and its aquatic larva. D In water with submerged plants rooted in lake bottom: (7) Water boatman or back swimmer (*Notonecta glauca*); (8) Great pond snail (*Limnaea stagnalis*); (9) Tufted duck (*Aythya fuligula*); (10) Flatworm (*Planaria* species); (11) Perch (*Perca fluviatilis*). (See also box right.)

Changes from Pond to Land

Most ponds and lakes have a limited life span simply because they are constantly being filled in by sediments, so they can be regarded as the early stage of a succession which leads toward the production of new land surface (*terrestrialization*). This process of infilling results partly from the local generation of organic matter by the free-floating plants and the plankton, and partly from the inorganic (and sometimes organic) matter which is being eroded into the lake from the surrounding catchment.

While the water is deep, more than about 2m (6.6ft), most plant life is free-floating, but as the sediment gradually rises some plants are able to invade which are rooted in the muds at the bottom but have floating leaves on the surface. Most familiar of such plants are the water lilies, which need very long leaf stalks if they inhabit fairly deep water. The presence of such plants has two effects on the course of succession. First, they increase the local productivity of the habitat and therefore add to the production of organic litter into the sediment; and secondly, they modify the flow pattern of the water, creating eddies and slowing currents so that more of the suspended sediment falls to the bottom. Both of these effects result in a faster rate of sedimentation and therefore a speeding up of the succession.

By the time that the water is only 1m (3.3ft) or so deep, other plants which have shoots projecting above the surface of the water can establish themselves. These are the emergent aquatics, and their invasion leads to great changes in the animal life of the lake. Many invertebrates, such as dragonflies, need emergent stalks from which to lay their eggs. New habitats and feeding sites are provided for certain snails, such as *Succinea*.

Areas of dense emergent plant life are called swamps, and these are particularly important for bird life. Many birds such as bitterns, coots and gallinules shelter in the reed beds and nest in them away from the sight of predators. All of these nest on platforms of dead vegetation just above the water surface, but others such as the Reed and Sedge warblers in Europe and the redwing in North America support their nests in the reed canopy and obtain most of their food from above the water surface.

Further development of this succession is dependent upon the continued rise of the "soil" surface until its summer level is actually above the water table. The habitat is now termed a fen and is likely to be invaded by such trees as willow, alder or Swamp cypress. Any continued successional change will depend on the rate of local peat accumulation. PDM organic matter (*detritus*) which is found in the sediments, much of it derived from the plankton. This material is made more worthwhile by the presence of bacteria on its surface. In addition, there are the large plants of the bottom. In fact these are not used much for food when alive, but provide a hard surface on which small algae and bacteria grow; these are grazed by herbivorous insects and mollusks. Usually the large plants only become food when they are dead and undergoing decomposition. The diversity in type and size of the grazers, detritivores and filter-feeders gives opportunities for a variety of invertebrate and vertebrate predators. Small water bodies usually only have invertebrate and amphibian predators, like newts and frogs; larger ones also contain fish.

Although large plants have little direct importance as food, they do have two other important functions. One is to provide highly essential shelter against predators for prey species. The presence of such shelter may often be the only reason for the persistence of a species. Plants have a further role in the life histories of many species, particularly insects, since they provide a means for larvae to leave the water to become free-flying adults. They also provide a means for females to enter to lay eggs, and a surface on which to deposit those eggs.

Flowing water is a very different habitat from static water. Most authorities distinguish two basic categories of flowing water—streams and rivers. Streams are cool, shallow, and often have a bottom of gravel, stones or boulders; whereas rivers are deeper, warmer and have a silty bottom.

Many rivers have recognizable sections, an upper course in hills or mountains which is essentially fast-flowing and stream-like, a transitional central section, and a slow-flowing, meandering final section in the flatter plain. Physical conditions are rather more variable than in lakes, mostly because of variations in flow. The strength of the current, the turbidity, the oxygen and nutrient content, and the temperature will all be affected by the volume of water in a river. The rate of flow depends upon the rainfall, and the differences between flood conditions following heavy rain and conditions following drought will be considerable. Differences in flow rate are often seasonal, many countries having a rainier season which is frequently the winter, while in high latitudes the spring thaw may constitute the period of greatest flow. In some cases rivers are wholly seasonal, drying up entirely at one season of the year. Whenever flow differences are great, the biota is found to be adapted accordingly. Life cycles are adapted so that drought periods are spent as resting stages, and periods of excessive flow may be similarly avoided. Rivers which have a more reliable flow, however, will have a more diverse flora and fauna.

The quantity of dissolved nutrients present in a river or stream will depend principally on the nature of the rocks of the area which the river drains. Streams originating on hard, impermeable rocks will contain little in the way of dissolved minerals, while those from softer, permeable rocks may contain an abundance of minerals. In general, soils derived from base-rich rocks are more productive, and so more plant material as well as more inorganic salts enter the streams from such areas, providing a basis for the differences in productivity of rivers.

Flow affects animal and plant distribution. Because the water moves quite fast and in one direction only, planktonic organisms are absent from all except the slowest-flowing rivers. The fauna is restricted to strong-swimming fish such as trout (nekton) and to bottom-gripping invertebrates such as stone flies, mayflies, limpets, leeches, and snails (benthic forms).

▶ **Spider catches fish.** The Raft spider (*Dolomedes fimbriatus*) is a voracious predator which normally hunts insects from leaves floating on the surface of ponds. However, it will go below the surface and has here captured a large minnow (*Phoxinus* species).

Both the nature of the bottom and its oxygen content depend on flow rates. In slow-moving water such as lowland rivers, small particles can settle and remain on the bottom, giving a muddy substratum. But as flow increases so too does the size of particle which can be moved by the water, until in very fast-moving upland streams only large boulders are stable. Rivers in the transitional section have alternating areas of shallow, fast-flowing water with a gravel or stony bottom and slower-flowing areas with a muddy bottom. There may be considerable variation in bottom type across the profile of deeper rivers. Thus a range of conditions will be available over a short distance, giving opportunities for a variety of animal and plant life.

Animals and plants are found over the whole range of water flow. Those inhabiting high-flow stretches have abundant oxygen and food, and their principal adaptations are for clinging to the stones. In fact, many stream animals have a preferred current speed which proves optimal for their success. For example, chironomid midge larvae prefer speeds of 80cm/second (32in/second) while the water shrimp *Gammarus* seeks currents of 15cm/second (6in/second) or less. Distribution patterns of animals on rocks, across and along watercourses reflect such preferences.

Animals living on or in the sediments, where flow is minimal (fish such as barbel, bivalve mollusks, worms and insects) have a different problem, namely that of obtaining sufficient oxygen. As a result, they have both morphological and physiological adaptations. Large mobile gills are a common feature of insects under such conditions.

The main food source for invertebrate animals, which predominate in streams, is detritus, decaying plant material and associated bacteria, much of it derived from land plants, particularly in wooded areas. Insects show elaborate modifications for acquiring this material. Evidence is accumulating that it is the bacteria on the detritus which is the principal source of food for many animals and not the detritus itself, the latter being relatively indigestible. In treeless areas, small surface algae may coat stones, rocks and fringing plants, forming a significant food source for grazing insects and mollusks. These animals form the food of both invertebrate and vertebrate predators, including birds like the dipper, which complete the food-web. In the lower reaches of rivers, although invertebrates remain abundant, vertebrates, especially fish, become more diverse and important in the community. In these areas, the water is so turbid that plant growth is limited to shallow areas and the banks. Eventually most rivers reach the sea, and in their lowest reaches the fauna and flora become modified by the effects of salinity on the freshwater species and the incursion of tolerant marine species upstream. RHE

INTERTIDAL MARINE

Problems of exposure to both water and air... Zonal patterns... Flora and fauna of rocky shores, and how they adapt... The food cycle... Life in tidal pools... Life on sandy and muddy shores...

THE intertidal environment is the zone between land and sea periodically exposed to the air by the tidal movement of water. The intertidal is the most physically exacting of all marine environments because it subjects marine organisms to periods of exposure to an alien medium, air. It is also washed by wind-created waves, which cause further problems for animals and plants. During the period when they are uncovered, marine organisms are liable to desiccation, exposure to damaging ultraviolet radiation, high and low temperatures, rainfall, and attack by land-based predators; and of course they must adjust their respiration to the changed conditions. Only adaptable species can survive and thrive in this environment. Despite these problems, animals and plants are abundant in the intertidal.

The time for which an organism is exposed to the air differs according to its position relative to the low- and high-tide levels, and the stress suffered increases with the height above the low-water mark. Only species highly adapted for a semiterrestrial life live at the top of the shore, where two-thirds or more of their time is spent out of water. Depending on its ability to cope with aerial conditions, each species has an upper limit on how far up the shore it can live. Most species are also restricted to a part of the beach on which they are better adapted for the prevailing conditions than their competitors. As a result, both plants and animals form bands or zones on intertidal beaches. These are more readily visible on **rocky shores** but are also a feature of sandy intertidal areas. On rocky shores, the major zones are so predictable and widespread that a universal zonation scheme has been proposed. Three major areas are readily recognisable. The first is a *sublittoral fringe*, characterized by species with very little tolerance of aerial conditions such as sea squirts, bristle worms and starfish, and in many areas recognizable as the upper limit of the kelps. The second is the *eulittoral*, the zone populated by animals and plants which vary in their tolerance of air but which depend upon the sea and mostly require daily immersion, such as wracks, barnacles, limpets and other snails. Commonly the upper limit of this zone is the upper limit of barnacles. Above this is the third area, the *littoral fringe*, populated principally by maritime species, that is terrestrial species adapted to saline conditions, such as lichens. It may also be invaded by some very tolerant marine species, including snails.

The most obvious plants on rocky shores in temperate and tropical latitudes are red, green and brown algae. Microscopic algae, the diatoms, cover all free surfaces—including other seaweeds and shelled animals—and are a major food source for grazing animals. Equally apparent are bigger seaweeds forming zones from the low-water mark upwards. Kelps or oarweeds are streamlined and flexible plants highly adapted to bend and give with the waves. They form dense forests at and below low water, but extend only marginally into the intertidal, being intolerant of desiccation. This part of the shore is also the major habitat of many beautiful and often delicate red, green and brown seaweeds which are unable to survive higher up the shore except in tide pools. The mid-shore in temperate latitudes is the home of some tolerant red and green algae, but is most characteristically populated by the hardy brown wracks. In Europe and North America these may form zones reaching as high as the high-water mark of neap tides. Lichens are also characteristic of many shores, often being mistaken for patches of tar because of their black, gelatinous appearance.

Barnacles, limpets, winkles, whelks, mussels, and starfish are the animals most obvious on rocky shores in cool latitudes and warm latitudes alike. In warm latitudes wormlike gastropods may also be found. These animals are often present in vast numbers and form zoned patterns, as do the plants. However, some species are found on the shore which are intolerant of aerial exposure, such as sea anemones, crabs and other crustacea, soft-bodied worms and sea urchins as well as sponges and sea squirts. The undersides of stable boulders, sheltered crevices and damp overhangs are thronged with a variety of such animals. Tide pools may form a refuge for some species.

Observation reveals that the numbers of species of plants and

animals decreases from low water upward as conditions become harsher. The intertidal is also a home for some terrestrial species whose problems are the reverse of marine organisms'. Many of these, which include insects and other arthropods, must seek trapped air bubbles at high tide.

The particular plants and animals to be found at any point depend on the extent of local water movement due to waves and currents. Few shores are so severely pounded by waves that no plants and animals are present, but it is a specialized few that are adapted to severe wave exposure. As wave severity lessens, different species become abundant, this trend being carried right through to areas of complete shelter and very little water movement. Different species are adapted to and most competitive at different degrees of exposure, resulting in diverse plant and animal communities. (See also p22–23.)

The plants and animals of rocky shores have evolved for a life partially spent in air in various ways. Plants become adapted to the intertidal by having thick, impermeable outer layers (which resist water loss), often dark in color (since excess ultraviolet light is a problem in the mid-shore and above) and by having low surface area, so that water loss is minimized. High-shore forms, on the other hand, have a considerable tolerance of water loss, which is necessary because they may be exposed to air for days on end. Some high-shore algae actually photosynthesize better out of water! Relative tolerance of desiccation determines the upper limits of seaweeds on shores, while competition determines the lower limits.

Tolerance of water loss also characterizes animals from the rocky intertidal, particularly those sedentary forms which cannot escape by hiding in crevices. A most remarkable example is the limpet, which may lose an astonishing 80 percent of body fluid before dying. Most successful high-shore forms have shells, which may be white and tall to reflect heat in warmer climates (mollusks and barnacles are examples). They resist water loss by clamping these shells tightly closed, or by clamping them firmly down onto the rocks. Some are able to survive

▲ **Squirting sea squirts.** View of a rocky shore in South Africa with sea squirts (or red bait) squirting water into the air.

► **Swift-footed rock crab** (*Leptograpsus variegatus*), exposed on the shoreline in Australia. Crabs make up a significant part of the fauna of the intertidal zone throughout the world, although they are intolerant of aerial exposure and hide under rocks or in pools at daytime low tide. Like many other scavengers and grazers they feed principally at the night time low tide.

the intertidal period by respiring without air (*anaerobically*), but many high-shore forms are adapted for breathing air. These have a tiny opening through which air can enter but from which only small quantities of water are lost. Water is also lost in waste products, and high-shore forms have evolved an ability to produce waste products containing little water.

The abundance of animals in the intertidal indicates that food must be readily available on rocky shores. The surface microalgae are eaten by winkles, limpets and other snails using the strong, many-toothed tongue (*radula*); and the phytoplankton brought in the water at high tide are eaten by the barnacles and mussels. The large algae are almost ignored as a food until they are broken off from the rocks, when they decay and become an important food source for the intertidal scavengers. The abundance of sedentary animals means that for adapted predators there is also plentiful food. Invertebrate predators such as dog whelks, starfish and crabs are most active in the low intertidal and in tide pools, and so their prey may only survive successfully above these limits.

Vertebrate predators such as fish and birds attack prey at all levels at high or low water as appropriate.

The appearance of a rocky shore changes through the seasons and from year to year as a result of mortality, seasonal influxes of short-lived species, the settlement of larvae, and competition between organisms. In temperate latitudes an early spring visit reveals spaces on the shore resulting from the removal of ill-attached or elderly individuals by winter gales. A visit in later spring and careful, close inspection reveals the presence of young individuals settled from the plankton. By summer the rock surface will be covered with growths of short-lived (*ephemeral*) algae and rapidly growing recruits to populations of longer-lived species. By late fall, the ephemerals will be gone, leaving the successful long-lived forms.

Tide pools form permanently wet habitats at all levels on beaches. On rocky shores they are frequently rich in algae, usually different from those of the surrounding rock and mostly from lower down on the shore. Physical conditions in tide pools, especially those rich in algae, are highly variable, the variation becoming greater with the height above low water. Temperature, salinity and acidity all fluctuate, but changes in oxygen level are often the most dramatic. By day, photosynthesis produces oxygen, leading to very high levels in an isolated pool,

◄▼ **Animals of a sandy shore** typical of northwestern Europe. Within the debris of the subterrestrial fringe or strand line: sand hoppers, for example (1) *Orchestia gammarella* and (2) *Talitrus saltator*. On the upper shore: (3) Masked crab (*Corystes cassivelaunus*); (4) a polychaete worm *Nephtys caeca*; and (5) an isopod *Eurydice pulchra*. On the mid-shore: (6) ragworm (*Nereis diversicolor*); (7) a polychaete worm *Amphitrite johnstoni*; (8) Sand mason (*Lanice conchilega*); and (9) Peacock worm (*Sabella pavonina*). On the lower shore and continuous shallow water: (10) Sea-potato (*Echinocardium cordatum*); (11) Common otter shell (*Lutraria lutraria*); (12) Pod razor shell (*Ensis siliqua*); (13) Thin tellin (*Tellina tenuis*); and (14) another tellin, *Tellina fabula*.

while at night the respiration of plants may cause oxygen levels to plunge very low. Only highly tolerant animals will be found in such pools. Lower down, the pools have a continuously changing population of creatures from the low shore and animals which live associated with algae.

Sandy and muddy shores often appear bare of life, for no large organisms are present at the surface at low tide. However, many species of burrowing mollusk (such as clams), polychaetes, crustaceans, sea urchins and other animals may be found by digging. As with rocky shores, there are distribution patterns with animals forming zones on the shore. There are many species confined to the lowest limits of the intertidal, incapable of surviving above these levels. Above them are found species characteristic of the intertidal. Many beaches, particularly steeply-sloping, coarse-grained, exposed beaches where wave action disturbs the sand a great deal, contain very few animals. Shores with a gentle slope, small-grained sand and with light to moderate wave activity are most likely to have a diverse fauna. When wave action is insignificant and the shore muddy, the range of species falls again, although the number of individuals of adapted organisms may be very high.

Above this zone on all shores is the subterrestrial fringe, populated by crustaceans, such as sand hoppers and crabs, which survive by burrowing into moist sand and by feeding on seaweeds thrown up by the tide. Although living at or above the tidal limits for part of the time, these animals are marine and may scavenge far down the shore at nighttime low tides.

Clear zonation patterns are often seen in sand and mud, but the reasons for them are not always clear. For sandy-shore animals, algae and bacteria attached to the sand grains form one food source, plankton another, and detritus brought by the tide a third, while for shoreline animals decaying rocky-shore weeds are a major food.

Sandy- and muddy-shore animals have different adaptations from those of rocky shores. As they live in the substrate they have no problems of desiccation, particularly since all are mobile. They must, however, be able to burrow, and adaptations for this are an obvious feature of all sand and mud dwellers. Many burrowing animals have well-developed respiratory pigments, can become quiescent, and can respire anaerobically in order to survive the periods when oxygen is not readily available.

RHE

MARINE

The largest and most diverse of the world's biomes... The major oceanic currents... Life in the seawater: algae, plankton, nekton... Life on the sea bottom... Filter and deposit feeders... Shallows and estuaries... The ecology of whales... Life in the bathyal zone... Animal adaptation below 2,000 meters... Life in areas of submarine volcanic activity...

THE marine is the largest of the world's habitat types. It covers 70 percent of the earth's surface, has an average depth of over 3km (about 2mi) and contains underwater chasms deeper than the highest mountains on land; these statistics give an impression of the size. The sea is immense. It is important both as the site in which life first arose and in which the greatest taxonomic range of animal types still live, and also because of the considerable influence it has on the climate of the continents. It is important, too, as a source of food; over 60 million tonnes of marine life is harvested annually by the world's fishermen—though this is a very low productivity considering the sea's vast extent.

Water is an excellent solvent, and sea water contains almost every chemical element in some quantity, and many compounds vital to animal and plant life in some abundance. Contrary to 19th-century views, oxygen is available at all depths, and so the sea is an ideal medium for life. Sea water has many other properties of importance to its inhabitants. These include transparency, which allows light penetration so that plants beneath the surface can photosynthesize—this is vital for the economy of the oceans. The ocean is characterized by low viscosity, which allows wind to obtain a frictional grip on the surface to create waves and currents as the water responds to those frictional forces, and by high density, which allows organisms to float without great effort within it.

The interaction between the atmosphere and the hydrosphere is continuous and global in extent. It results in the generation of the world's wind and major ocean current patterns. The whole of the world's oceans are in continuous three-dimensional movement. Some movement is obvious: waves and currents due to the winds and the tides are familiar, and they are indeed important in determining the bottom type and the animal and plant distribution in shallow water, but their action is very minor compared with either surface or deep ocean currents.

The surface currents, those of the top 50m (165ft) of the sea, are caused by the frictional action of wind on the water surface, and the patterns of flow result from the interaction of this with the rotation of the earth on its axis. In Northern Hemisphere oceans, for example, there is a clockwise set of surface currents driven by the trade winds and prevailing westerlies of higher latitudes. Water moved in this way must be replaced, and thus there must be counter-currents. Similar patterns are found in all the world's oceans. At certain points on the earth's surface physical conditions such as high salinity and low temperature combine to make the water dense. This water sinks to the bottom, and becomes part of the general deep circulation, flowing along the bottom and mingling with overlying water

masses. As a result of these and other mechanisms, the oceans are made up of a number of discrete bodies of water flowing in various directions. Many of these are enormous in volume, large enough to have distinct animal communities associated with them, and oceanographers can often identify the origin of a sample of water from the animals it contains. However, because water masses are continuous there are also some species which may be found all over the globe.

The flows are sometimes quite rapid; the major current known as the Gulf Stream flows at 2m/second (7ft/second) off the eastern coast of the USA. Most, however, are slower; but even in the deep sea, flows of 0.2m/second (0.7ft/second) have been recorded. The flows have the important functions of circulating oxygen, nutrients, plants, spores and larvae as well as adult pelagic animals such as turtles, and also provide an orientation and navigation aid for many migratory species.

Two main habitats exist, the water itself (the *pelagic* habitat) and the bottom (the *benthic* habitat), each populated by a diversity of animals and plants. They extend, in the case of animals at least, from the intertidal to the greatest depths. Three other habitats are also distinguishable: *estuarine*, *bathyal* and *abyssal*.

The **pelagic habitat** is divided into three parts: that in which sufficient light is present to allow photosynthesis and thus the production of plant tissue, that part in which some light is present, and the deep, dark areas. Light penetrates water relatively poorly, and thus photosynthesis even in the most transparent of water takes place only in the top 100m (300ft) or so, and light does not penetrate at all below 1,000m (3,300ft). In water containing suspended material, light penetration may only be a few meters. Thus the productive part of the sea, where solar energy is used to manufacture plant tissue, upon which ultimately almost all marine creatures depend, is only a very thin skin.

The actual plants of the sea are of several kinds; most important are the microscopic planktonic species inhabiting the surface waters. These are the unicellular algae called diatoms, only 20–200μm (0.000002in–0.00002in) long. Together with others called silicoflagellates they are responsible for 90 percent of the plant production of the sea, which in total represents only about 25–35 percent of the total primary production of the earth. Diatoms predominate in temperate latitudes, silicoflagellates in warmer areas. The remaining 10 percent of plant productivity comes from much larger red, brown and green multicellular seaweeds such as wracks and kelps, as well as marine grasses and mangroves attached to the bottom in water shallow enough to allow photosynthesis and growth.

Production by plants is unequally distributed over the oceans. High production only occurs in areas where the essential nutrient elements, nitrogen and phosphorus, are present in sufficient quantities. Such areas include shallow waters of continental shelves where storms keep the water mixed and there is runoff from the land, and places where nutrient-rich deep water wells up to the surface. Much of the deep water in tropical latitudes has low levels of essential nutrients and is relatively unproductive.

◀ **Fish and coral**—a coral reef scene in the Red Sea. Coral reefs are very complex ecosystems containing many species.

◄ **Poking its head through an ice floe,** a Weddell seal (*Leptonychotes weddelli*) comes up for air. Weddell seals rarely come ashore and mostly give birth on ice floes. They feed mainly on fish and cephalopods, and some bottom invertebrates (secured by deep-diving). Like most seals, their lives and habits when living in the open ocean remain a mystery.

► **Ocean drifter**—a moon jellyfish, *Aurelia aurita*, hangs in the water just below the sea surface off Sweden. Although jellyfish are able to move of their own accord, mainly to stop them sinking, they are at the mercy of tides and currents.

► **Life on the sea floor** BELOW. This mixture of sea urchins, brittle stars, crabs, scallop and seaweed emphasizes that the benthic habitat may contain a great variety of life. This photo was taken in the saltwater Loch Duick in western Scotland.

In the tropics production, although low, is continuous; in temperate and higher latitudes it is seasonal with peaks in early spring and in fall. The life patterns of many animals are arranged to take maximum advantage of this production—thus for example many species with larvae which feed on phytoplankton reproduce in the spring. Production is low in winter because of low light and temperature levels, and low in summer because nutrients are not available. This is partly because the nutrients have been absorbed and used to make animal tissue, and partly because in summer the warm surface layers of the sea become separated from the deeper cooler waters, and the nutrients in the surface waters rapidly become depleted and cannot be replaced. It is when this so-called *thermocline* breaks down that the second production peak occurs in the fall. Such a thermocline does not appear in vigorously moving water, and so is not found in southern temperate latitudes where strong winds are normal. The southern ocean is therefore continuously productive while light is available, and supports huge populations of animal plankton as well as sea birds, seals, fish and whales. Until recently it was the main center of whaling, and is now the main area for exploitation of a form of shrimp called krill, the main food of the few remaining whales.

The animals of the pelagic zone are of two types, plankton and nekton. Planktonic animals are drifters—with limited powers of movement, their distribution is largely the result of water movements. Nekton are active swimmers, usually larger than plankton, and their distribution is largely the result of their own movements.

Plankton include representatives of most major groups, but the crustacea, and in particular one group of these, the copepoda, predominate. In addition, the larvae of many bottom-dwelling creatures may be temporarily planktonic, feeding on the minute plants and acting as a distributive stage. Some plankton groups such as arrow worms, sea gooseberries and true worms are exclusively carnivorous, but the copepoda are predominantly herbivorous— they are the main grazers of the phytoplankton, and also the main food resource of carnivorous zooplankton and of smaller nekton.

Planktonic animals live in an environment in which concealment is both difficult and necessary, and most animals show adaptations to enable them to avoid their predators. Many, like jellyfish, are virtually transparent, others, such as shrimp-like crustacea, are disguised by disruptive color patterns, or (in the case of deeper-water forms) by light patterns. In addition, many have protective spines and lumps to put off predators. Even this may be inadequate to prevent predation, and it is very likely that the common tendency to undertake vertical migration may, as well as placing the migrator near its food, also reduce predation. Certainly the pattern of migration of most species, which brings them nearer the surface during darkness hours and leaves them deep down in daylight hours, suggests that they are avoiding situations where they can be more easily seen.

Pelagic organisms have another common problem: how not to sink. Animals can usually swim, but this is wasteful of energy, and both animals and plants have found ways of offsetting gravity. These include the elaboration of gelatinous tissues, as in jellyfish, storing light oil droplets, as in shrimps, gas bladders and chambers, as in fish, replacement of heavy ions in the cell fluids with light ones, and developing shapes which reduce rates of sinking.

Nekton are actively swimming forms, such as squid, fish and vertebrates. In addition to being much larger than planktonic forms and therefore able to move faster, nektonic species are highly adapted for efficient swimming and for detecting the presence of prey or predator—they have large eyes and other well-developed sensory systems. They protect themselves against predators by disguising color patterns, counter-shading and the like. Some are extremely large, a tribute to the nutritional quality and abundance of the zooplankton. Nektonic species are almost exclusively carnivorous, feeding on plankton, other nektonic forms or animals of the benthos. Most species commercially important to man are nektonic.

In the **benthic habitat**, animals are found in and on the bottom, from positions even above high-tide level to the greatest ocean depths. Plants, however, with the exception of microbes, are found only in shallow waters, where light levels allow their growth. The bottom itself may be solid rock, coral or particulate material ranging from boulders to the fine mud and ooze of the deep sea, and each substrate type will have benthic species adapted to it and characteristic of it. In addition, because of their preferences for particular benthic species, it is common to find fish associated with particular bottom types. The particle size of the bottom is directly correlated with water movement: areas of high flow will have a substratum of coarse sand or gravel, if not bare rock, while where flow is gentle, mud or silt will form the bottom.

The different substrates have communities dominated by particular feeding types. Thus hard surfaces and coarse sediments below low-tide level are dominated by filter-feeding animals and their predators. Filter feeders are animals which actively draw currents of water, either through a sieve within the body or through a sieve structure extended into a water current. Classic filter feeders are mussels, sponges, sea squirts and barnacles, as well as fan worms and brittlestars. In sand, filter feeders will be equaled in abundance by deposit feeders, and it is these which predominate in all fine sediments, often completely excluding filter feeders. Deposit feeders actually process the sediment in some way, and many live buried in it. Their food is the organic detritus in the sediment and associated bacteria; although not as nutritious as the plankton food of filter feeders, this is available reliably all year around. Most animal groups have representatives capable of deposit feeding, but worms, mollusks and crustacea predominate. The feeding activities of deposit feeders result in complete working over of the sediments, and can change the nature of the soil.

It has been found that every substrate type has a characteristic community associated with it, and since the evolutionary forces dictating which types are successful will be the same the world over, the community patterns are also worldwide. In some cases it is the same genus of animal which is found the world over.

The productivity of benthic communities depends on the food available, and is related to local primary productivity. Shallow water thus usually has the highest productivity; and, in general, the biomass of animals declines with depth and distance from land. Thus very deep water is populated by small numbers of small animals. The major fisheries for bottom-dwelling fish are thus based on shallow-water areas such as the North Sea and the Gulf of Siam.

Estuaries are a small but extremely significant component of the marine environment. They are the points at which rivers empty into the sea, and are characterized by having water of low and fluctuating salinity. Most rivers also deliver significant quantities of mud, detritus and plant nutrients to the sea. Estuarine animals are among the most adaptable of marine creatures, being capable of withstanding salinity change and coping with the considerable problems of living in mud. For animals capable of exploiting this environment there is, however, a vast amount of food available, and the biomass and productivity of animals in estuaries is often very high.

The importance of estuaries lies in the fate of this production. The shallow waters of estuaries are used as nursery grounds by many economically important species of fish, the young fish finding abundant food for rapid growth among the plethora of small invertebrates. The high productivity of estuaries also results in their being ideal for the growth of shellfish for human consumption. Much of the high output of estuaries is exported to the neighboring sea area, significantly increasing the production possible there.

Of course, estuaries are also frequently sites of heavy human

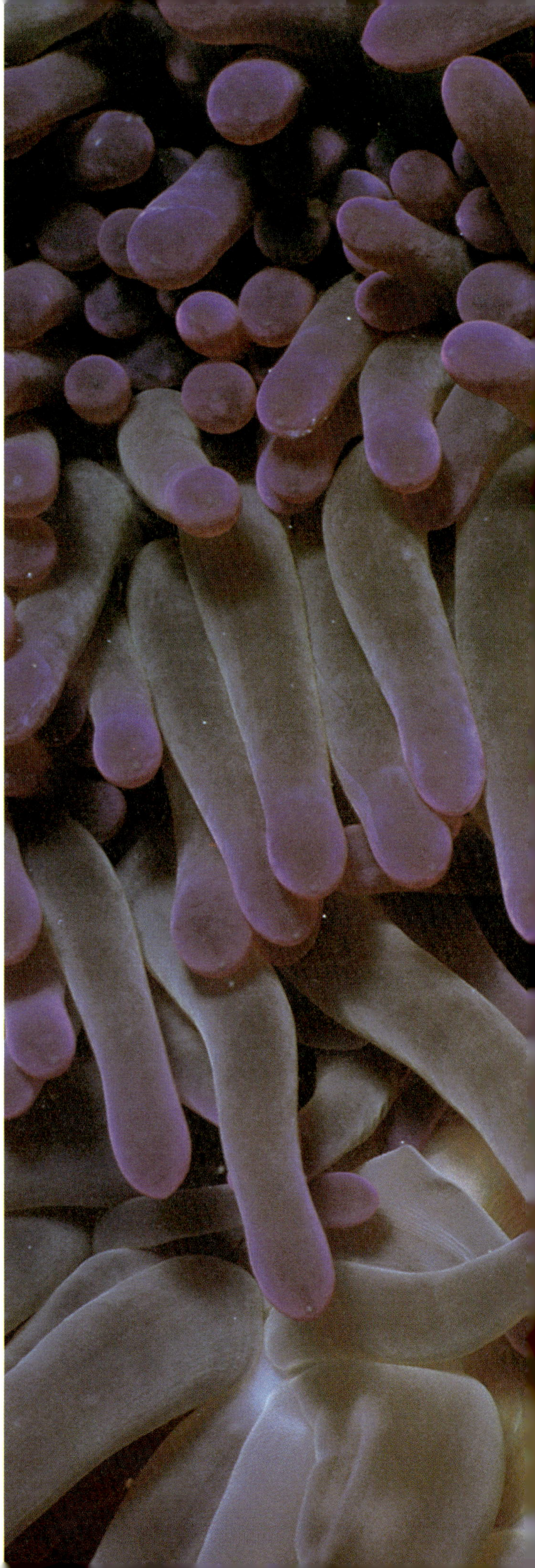

► **Kaleidoscope of tentacles**—a close-up of the tentacles and mouth area of a sea anemone (*Stoichactis* species), photographed on the Great Barrier Reef, Australia. Sea anemones are a conspicuous part of benthic fauna throughout the world.

activity, and are the route by which much pollution—including heavy metals and oil spillages—enters the marine system. Fortunately there is little evidence at present of permanent or long-term damage being done to marine habitats by pollutants.

Beneath the well-lit surface layers of the sea, where photosynthesis is possible, lies a twilight zone in which light diminishes and pressure increases with depth. This is the **bathyal zone**, the home of some of the most bizarre of marine creatures. It stretches from 200m (660ft), where light is quite apparent, through 1,000m (3,300ft), where all light is gone, to 2,000m (6,600ft), where it gives way to the abyssal zone. The problems of living in this habitat are few. The main one is not pressure, which appears to have little effect on the physiology of animals, but the very low availability of food. The only food for pelagic animals is plankton, waste materials from surface-dwelling animals, or fellow pelagic creatures. Food from all three sources diminishes with depth. The majority of animals are predatory, and the most spectacular adaptations are found in those species which must be highly efficient as predators and yet avoid falling prey themselves to others.

Since food is scarce, and long periods may separate opportunities for predation, deep-sea species must reduce their energy use between meals, have highly efficient methods of attacking and holding their prey, and the capacity to store and digest large meals. Bathyal fish therefore show most of the following features: small size; reduction of muscles and other heavy organs; extensive elaboration of the sense organs for prey and enemy detection; very extensible jaws; long, recurved, sharp teeth; a massive gut; and a color pattern designed to provide maximum camouflage, and very often including bioluminescent organs of amazing sophistication. Not all bathyal species wait for food to come to them. Some species migrate very considerable distances vertically to feed near the surface at night.

The bottom fauna is diverse at these depths but dominated by deposit feeders able to process the poor-quality, indigestible material which falls from above. Many are dependent upon bacteria for survival. The quantity of bottom fauna decreases with depth. The individuals are mostly smaller than shallow-water forms, they grow more slowly, reproduce only very infrequently, and may live for many years.

The **abyssal zones** are the parts of the oceans exceeding 2,000m (6,600ft) in depth. They are cold, dark and food-deficient, and subject to great pressure, but these conditions are very constant. The bottom is for the most part covered by ooze or mud, and water flow is low though usually perceptible (0–160mm/second, 0–6in/second). Near land, the bottom sediments are of land-derived materials, but most oozes are made up of the skeletons of oceanic protozoa. At depths of 4,000m (13,000ft) and over the bottom is always a red clay.

A remarkable variety of species is found on or near the bottom in the abyssal depths, but diversity and abundance do decline with depth and some groups disappear. Species well adapted for processing sediments, such as worms, bivalves and sea cucumbers, predominate, while fish are scarce.

Abyssal animals are highly adapted to an environment in which the principal problem is lack of food. They need to be highly efficient and to reduce in size all non-essential organs. They are thus mostly small and frail in appearance, white or grayish in color, and blind. Most compensate for blindness by developing organs sensitive to touch and chemicals in the water. All have specialized guts to enable the maximum nutritional benefit to be obtained from food. Detritus feeders living on and in the bottom will feed on the undigested remnants of materials produced at the surface and the skeletal remains and excreta of animals from the overlying water. Often only bacteria can break down this material, and thus deep-sea animals often culture bacteria in the gut. Highly sensitive organs of smell and touch characterize the scavenging and predatory animals of this habitat, as also do highly distensible guts. Typical animals of the abyss, like those of the bathyal biome, grow slowly and reproduce infrequently, produce few young and often give them brood protection.

Most of the abyss is a muddy plain, but at certain points in the sea floor, where the continents are separating from each other, the earth's crust is thin and submarine volcanic activity is evident. Geologists using submersibles and other sophisticated equipment have found submarine lava flows and gushers of hot, sulfur-rich water at various places in the Pacific. Around these hydrothermal vents, and even in them, a fascinating new community of animals has been found, unique in the world's oceans. The feature which stands out immediately as different from the fauna of the rest of the deep sea is the size and abundance of the animals. Some are huge even by shallow-water standards. Either they must be very old, or growth rates and hence food supplies are very different here. The most spectacular examples are gutless worms 100mm (4in) in diameter and up to 3m (10ft) long living in intertwined masses in white tubes. Equally unusual are enormous bivalves 250mm (10in) long and very abundant. Intermingled with them may be seen limpets, bristle worms, crabs, sea anemones, fish and other creatures, mostly large. Until very recently, all of these creatures were unknown to science.

Physiological experiment has shown the animals to have metabolism and growth very like shallow-water forms, which means that they must have an enormous and reliable food source, a situation unique in the deep sea. Two kinds of food are available, one derived ultimately from conversion of the sun's energy, the other uniquely from bacterial chemosynthesis. The first is suspended material, derived from surface activity, brought to the animals around the vent by the circulation created by the release of the hot water. Calculations have shown that an average vent would drag into the area around it as much as 30kg (66lb) of food a day. More interesting, however, was the discovery that bacteria present at the vents are capable of using the sulfurous chemicals as a basis for the synthesis of organic material. These bacteria are present in enormous numbers and form a major food source. Some combination of these two food sources fuels these unique "islands" of high productivity. Unusually for the deep sea, the hydrothermal-vent animals may produce very many tiny larvae. The reason for this difference from other deep-sea species is probably that vent systems have a limited life (perhaps only 50 years) and rapid distribution of many larvae is essential, if new vents are to be colonized. RHE

▼▶ **Wonders of the deep**—a selection of fish species that occur in the bathyal and abyssal zones of the sea. (1) Threadtailed grenadier (*Macrourus filicauda*). (2) An angler fish *Melanocoetus johnsoni*. (3) Whalefish (*Cetomimus indagator*). (4) A loosejaw *Malacosteus niger*. (5) Viperfish (*Chauliodus sloani*). (6) Garman's hatchet fish (*Argyropelecus affinis*).

Leviathans of the Oceans

The ecology of whales

Whales are the largest mammals, and, indeed, the largest animals of any kind, to inhabit the earth. Whales belong to the order Cetacea, which is divided into two suborders: the Odontoceti or toothed whales include the porpoises, dolphins and Sperm whale; while the Mysticeti or baleen whales include that largest of all creatures, the Blue whale. Although there are about 66 species of toothed whales, there are only 10 species of baleen whales in three families.

The family Eschrichtiidae includes only one species, the Pacific Gray whale (an Atlantic Gray whale became extinct about 400 years ago, apparently due to whaling activities). Gray whales are coastal and shallow-water creatures. They feed mainly in the shallow parts of the Bering Sea in summer, and they migrate along the northwest Pacific coast during fall and spring, to and from the shallow lagoons of Baja California.

Gray whales feed mainly on the bottom, sucking tube-dwelling invertebrates called amphipods into their mouths, and expelling the mud and water which enters the mouth during feeding through short bristly baleen. Gray whales are fusiform in shape, tapering toward each end, but they are not built for speed. They are dark gray, with mottled light and dark patches of skin, and they blend well into the usually murky waters nearshore. Perhaps because they habitually meet at certain traditional geographic places at both ends of their migrations, and because they travel a rather set route near the shore during that migration, they have not developed some of the elaborate acoustic repertoires found in the other two families of whales.

The family Balaenidae consists primarily of the Northern and Southern races of the Right whale, and the Bowhead whale (a third species, the Pygmy right whale, has not been studied, and very little is known about its habits). These whales also migrate between feeding and breeding grounds, but they do so over deep water, and they stay in acoustic contact much of the time. They also feed by opening their mouths wide and filtering small prey—such as calanoid copepods and euphausiids—into mouths containing thin, fine baleen. The baleen of a 16m (52ft) Bowhead whale may be as long as 4.5m (15ft), and the mouth may open to almost one-third the length of the body as the whale slowly moves through huge clouds of small prey. Like the Gray whale, these animals are not built for speed. Right whales and Bowhead whales at times cooperate in gathering food, by swimming side by side and in staggered or echelon formation in order to keep their prey from escaping to the side.

The third family of baleen whales consists of the Balaenopteridae, or rorquals. There are six species, ranging in size from the Minke whale (about 8m/26ft long) to the Blue whale (up to 30m/100ft long). All of them are extremely sleek and streamlined, with small thin pectoral flippers and a small dorsal fin (the Right whale, Bowhead whale and Gray whale do not have a dorsal fin). Rorquals are obviously built for speed, and the flippers and dorsal fin probably serve to stabilize these animals. They all have many throat grooves extending from the chin to near the navel, and they can open their mouths rapidly and engulf huge quantities of water and prey due to the extension of the throat grooves, much like the extension of the pouch of a pelican while it is feeding. The rorquals are lunge feeders which actively and rapidly swim through larger

▲ **Sifting the seas**—a pair of Humpback whales (*Megaptera novaeangliae*) feeding at the surface.

▼ **Baleen whale feeding techniques.** (1) Right whales (*Balaena glacialis*) open their mouths wide to allow food items to be filtered out by their baleen plates as they move forward. (2) Rorquals fill their mouths with water and then close it, forcing out the water and leaving the food in their mouths.

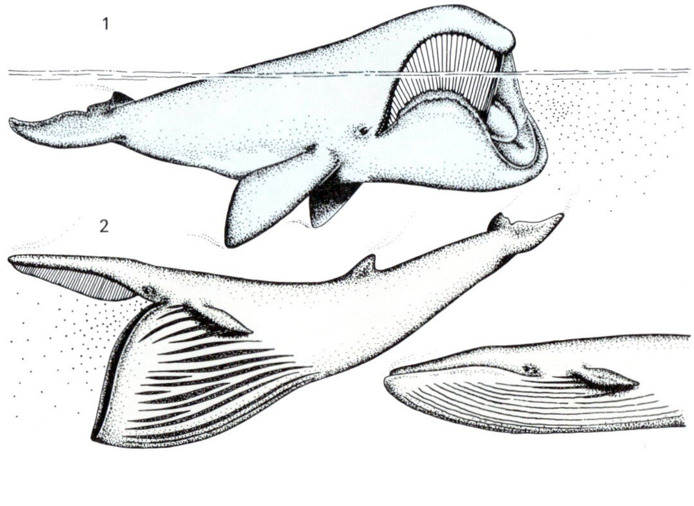

prey—such as krill, schools of herring, capelin, or other small schooling fishes. Although feeding often occurs over highly productive banks and near oceanic upwellings, rorquals traverse the deep ocean while migrating, and at least the Fin and Blue whales have developed low-frequency sounds which are so loud that they travel for dozens of kilometers. It has even been suggested that such low-frequency sounds may enter oceanic sound channels created by particular pressure-salinity-temperature gradients (the so-called SOFAR channels), and that their sounds may travel for hundreds of kilometers.

The differences in morphology and feeding in the three mysticete families are dramatic. The Gray whale is a stubby, generally slow animal which feeds on prey that moves little. The Right and Bowhead whales are more active animals which feed on huge clouds of moving prey. But it is the rorquals who appear most agile, and they also feed on prey which is itself highly maneuverable. Although there are exceptions in particular populations, it is also fair to say that the Gray, Right and Bowhead whales generally inhabit coastal or shallow water, while the exceptionally sleek rorquals are open-ocean pelagic species.

Toothed whales also show remarkably variant adaptations depending on habitat. One example stands out: the coastal species, such as Bottle-nosed dolphins, usually occur in small groups numbering at most several dozen animals. The pelagic species, such as Common dolphins, occur in groups numbering up to several thousand individuals. Apparently this difference has much to do with foraging once again. While small groups near shore often cooperate in feeding on small schools of coastal-dwelling prey, oceanic species cover a very large swath of sea as they search for prey, and they communicate throughout the school once prey is found. There may be another reason for the difference in group size between coastal and pelagic forms: the coastal animals have particular home ranges, and groups may segregate by age and sex. But because of the defined ranges, they know where to find each other for social activities and reproduction. In the large expanse of the ocean, where pelagic dolphins may travel for hundreds of kilometers in a few days, such separation would be inefficient, and they instead segregate by age and sex only within the confines of the school. Predation pressure probably plays a role here as well, for just as the collective sensory capabilities of many animals makes it easier for them to find food, so it alerts them to danger more readily.

BW

ISLANDS

Effects of isolation on animal species. . . Lack of predators. . . Unusual plants. . . Patterns of fauna and flora. . . National Parks resemble islands ecologically. . .

ISLANDS have always been of particular interest to biologists concerned with evolution. Isolated by the surrounding ocean, each island contains only a limited variety of organisms. It is therefore much easier to study how these interact with one another in affecting the pathways that their evolutionary changes may follow. The island environment is also sufficiently different from the mainland environment to allow enlightening comparisons to be made between evolution in the different circumstances. Finally, since no two islands are identical, comparisons between different islands are now leading to interesting attempts to produce general theories about the interactions between rates of colonization and extinction and the sizes and locations of islands.

It is not surprising that it was the study of island life that led both Charles Darwin and Alfred Wallace to the theory of evolution by natural selection. On the continents, each type of organism is accompanied by a great variety of competitors, and each has usually to become very specialist in order to coexist with them. Because islands receive only a limited sample of the continental animals and plants, the successful immigrant is freed from this competition and may find that it is able to colonize ways of life that are normally closed to it. So, for example, Darwin noticed that in the Galapagos Islands in the Pacific Ocean there were many types of finch. Although, on the mainland, finches all ate seeds, in the Galapagos Islands they had evolved into a variety of birds with different sizes and shapes of beak, eating different types of food. Some had evolved heavy, parrot-like beaks and cracked nuts to extract the contents, while others had evolved more delicate beaks like those of flycatchers, and ate insects. So, on different islands, evolution by natural selection adapts each organism to the opportunities of its environment.

Islands frequently lack large predatory carnivorous animals. This is mainly because of the difficulties they experience in colonizing remote sites surrounded by sea. In the absence of predators, larger types of flightless bird often evolve on islands; for example, the kiwi and moa of New Zealand, and the dodo of Mauritius (although, sadly, both the moa and the dodo became extinct because of predation by man).

Similarly, because the larger seeds of most trees are not well adapted to crossing wide stretches of water, islands often lack normal types of tree. As a result, other types of plant may evolve into trees on islands. For example, plants related to the little tarweeds of North America have colonized the Hawaiian Islands, where they have evolved into a number of shrubs and trees 3–8m (10–26ft) tall, belonging to the Hawaiian genus *Dubautia*. Their relatives the silverswords (*Argyroxiphium*), also a Hawaiian genus, are notable for the fact that they have colonized a wide variety of habitats, from boglands and rain forests to bare, cinder-covered lava peaks.

The comparatively small size of islands has also affected the size of the animals that inhabit them. The chances of extinction

◂ **Wallowing tortoises**—Giant tortoises (*Geochelone elephantophus vandenburghi*) in pools on the floor of Alcedor Crater, Isabela, Galapagos Islands. Giant tortoises of a different species also occur on the Seychelles. Several subspecies of both species inhabit the small islands of the two archipelagos, but some have died out or are threatened with extinction. These island tortoises originally had no natural enemies, and as well as growing to a great size, their shell has become relatively thin.

◂ **Flightless rail** BELOW of New Zealand. The takahe (*Notornis mantelli*) is unique to New Zealand and, as is the case on many islands that lack natural predators, it has abandoned flight. It was once numerous, but in this century was thought to be extinct until its dramatic rediscovery in the remote, high, tussock-grassland in 1948. Only around 300 birds exist in the wild, so it remains an endangered species.

are lower if there is a larger number of smaller animals, rather than a smaller number of larger ones. That is why we find, for example, fossil pygmy elephants on islands in the Mediterranean and the East Indies. Some animals, however, have evolved a larger body-size than is usual for their group, owing to the absence of predators. The dodo (a type of pigeon that once lived on the island of Mauritius) is an example of this.

Apart from these particular examples of evolution in the isolation of islands, biologists also try to discern general rules that may govern the diversity of island faunas and floras. The most detailed attempt at such a synthesis of data is the theory of island biogeography put forward by the American ecologists Robert MacArthur and Edward Wilson in 1967. They suggest that there is likely to be an approximate equilibrium level in the number of species in any island. This equilibrium results from a balance between the rate at which new species colonize the island, and the rate at which existing island species become extinct. When an island first becomes available for colonization, the rate of immigration of new species will initially be very high. Subsequently it will diminish, because most of the species that can easily colonize the island have already done so, and only the occasional lucky chance provides an opportunity for a new species. At the same time the rate of extinction, which was low at first, will rise. This is both because there are progressively more and more species, so that the numbers at risk are correspondingly greater, and also partly because the increasing intensity of competition as new species arrive makes it more likely that some of the species will become extinct.

MacArthur and Wilson realized that other factors will also affect the final size of the island fauna and flora. For example, larger islands are likely to provide a greater variety of habitats, and will therefore have a correspondingly greater variety of species. Similarly, since islands depend on neighboring continents as the source of their immigrants, islands that are closer to the mainland will receive more colonists than those that are further away, and will also be richer if that source itself contains many species than if it has a less diverse fauna and flora. However, the interactions between living organisms and their environment are so complex that it seems unlikely that they can be reduced to mere numerical predictions. This is particularly true of the environments today, both because they have been greatly affected by the changing climate and sea levels of the ice age of the last two million years, and also because 20th-century man himself has been changing the world at an unprecedented rate.

Although we are accustomed to think of islands only as land areas surrounded by water, any isolated environment is effectively an island, whether it is a mountain peak or a cave. So, too, are the protected areas designated as National Parks so that their (often rare) inhabitants can be conserved. The theory of island biogeography suggests that such parks should be as large as possible and should be close together or close to other areas from which they may gain natural colonists to sustain their numbers.

CBC

Man and Nature

MAN is an animal with ecological requirements just like any other. Human beings have physical limits of tolerance and optimum levels for survival; they also need a source of energy and mineral nutrients. But humans are peculiar in that they have the capacity to modify global ecosystems on a very considerable scale in order to redirect the products of those systems into the needs of human society.

No other organism has produced such an impact on the physical, chemical and biological constitution of the earth. For this reason it is necessary to look carefully at the relationship between man and nature through the eyes of the ecologist. In this way it becomes possible to locate those areas in which the stresses created by human beings are becoming intolerable and are threatening the future of certain species and habitats. It is also the most hopeful means of predicting the outcome of some of these harmful activities and may supply the answers to a number of the issues which are creating the current environmental dilemma.

◄ **Burning the earth's lungs**—clearing tropical rain forest in the Amazon. Rain forests are being cleared at such an alarming rate that there are fears that the atmosphere may be affected.

AGROBIOMES

Man and the environment. . . The history of agriculture. . . Simplification of the foodweb. . . Fertilization. . . Energy input and output. . . Economic considerations. . . Problems of disease. . .

THE influence of humankind on the ecology of our planet has become so marked that he has created distinctive environments of his own. One of these, the agricultural habitat is the major productive resource base upon which he is dependent. The immense impact of man is not surprising when one considers that he represents 4 percent of the animal biomass of the planet. Of the earth's total primary production, 7.7 percent results from human agriculture. Of the grazing which takes place on the earth's land masses, 14.5 percent is due to the activity of man's domestic animals. In theory the earth should be able to support man quite easily—mankind's total energy requirement is equivalent to only about 10 percent of the earth's primary production. But the demand for energy and food is concentrated in certain regions, and this, together with local climatic limitations, results in large disparities in their availability.

The agricultural habitat has had many independent origins in different parts of the world. The domestication of plants and animals was essentially a process in which man entered into a kind of symbiotic relationship with them. He increasingly depended on them for food, but in return he gave them protection from pests and predators. In the case of animals, he modified the vegetation in order to provide them with better grazing. By clearing woodland and burning scrub, he increased the productivity of palatable plants within the reach of his favored species. Indeed, even before true domestication had taken place, Middle Stone Age people in northern Europe are believed to have managed their environment in this way to improve conditions for wild Red deer and so to increase their populations and their reliability as a food resource.

For the plants, man assured their growth and reproduction by removing competitor plants, the weeds, by increasing the water supply through irrigation, and by providing nutrients in the form of animal manure. Part of the resulting productivity, of course, is consumed by man, but in return a proportion of the seed is conserved for planting the following year; the plant is assured of a future.

So areas of ground were selected for particular types of management according to the resources available. Arable agriculture was most successful in the rich alluvial soils in valleys, but such sites would have been the most difficult to clear of forest, so often less advantageous areas were used and were subsequently abandoned as their soils became exhausted. The upland areas were often easier to clear of forest, but less suitable for arable crops, so they were frequently used for grazing.

The development of agriculture, however, is accompanied by certain risks. Man becomes more vulnerable in some respects, for any circumstance which is adverse to his domesticated animals and plants may do him harm. Whereas hunters may inhabit extremely harsh environments such as the tundra, the agriculturalist is limited by the failure of his crops and domesticated animals.

► **Man's impact on the environment** through agriculture has changed the habitat in many ways. His grazing animals, such as the cattle and donkeys shown BELOW, contribute 14.5 percent to the grazing by all animals. Also RIGHT arable crops, such as this oil seed rape, are often in monocultures which exclude most other life-forms.

The agricultural biome is very simple in comparison with most other biomes. Man has taken great pains to eliminate all the components of the system which are not to his direct advantage. In this way, the diversity of species present (both plant and animal) is reduced to a minimum. In an arable system, the only organisms tolerated apart from the plant crop itself are certain soil microbes such as nitrifying and nitrogen-fixing bacteria and various arthropod detritivores and microbial decomposers which enhance the movement of nutrients from an organic to a usable inorganic form. This simplification of the ecosystem means that foodwebs become reduced to virtually linear patterns in which the flow of energy moves directly from plant to man in the case of arable systems and from a wider resource of plants via grazer to man in the case of pastoral systems.

Modern agricultural ecosystems in developed, mechanized countries have taken on a novel energy-flow pattern. Here, in addition to the solar energy which is available to plants and enables them to photosynthesize, man injects further energy in order to enhance the plants' development and growth. The ground is prepared using tractors which burn petrol, so extra energy is being used to give the seed a better chance of germination and survival. Pesticides and herbicides may be used to protect the young plants, and fungicides are often applied to the seed before sowing. These compounds are produced by industrial processes which are very expensive in energy. Fertilizers are applied to enhance growth, and whereas in the past animal dung was used for this purpose, chemical fertilizers are now more frequently used. One of the important elements in these fertilizers is nitrogen, which is a vital component of all animals and plants, and this is produced by fixation from the atmosphere in industrial processes which are extremely costly in energy. Further energy subsidies may be made to the crop in its irrigation, harvesting, cleaning, packing and transportation.

When all the energy, mainly in the form of fossil fuels, which man puts into the management of the agrobiome is considered, the question may be asked: is it possible to harvest as much or more energy in the form of food? This energy balance-sheet

approach can be illustrated by reference to the potato-growing process. The energy needed to produce a crop of ten tonnes of potatoes in the British Isles amounts to about 20,500 kilojoules (kJ). Of this, the main input is from machinery fuel, including tractors (9,800kJ) and fertilizers (6,100kJ). Of the ten tonnes of potatoes, 10 percent will be used for propagation in the next season and about 10 percent will be lost because of harvesting damage or pests. There is then the cost of transporting the salable potatoes, about 2,100kJ, and finally the losses due to peeling (about 27 percent) must be taken into account. The final energy output which is consumed by man amounts to about 26,000kJ.

Comparing this energy output with the energy invested in the crop, which totals 22,600kJ, it can be seen that there is only a small energy profit. The output to input ratio is in fact 1.15. So the agricultural production of potatoes is, in effect, a process whereby fossil fuels are made into edible starch.

Some crops grown under intensive cultivation are more

efficient than potatoes in terms of energy input versus output. For example, wheat has an output:input ratio of 2.2 and maize 2.8. But these figures refer to the grain crop only. If the remainder of the above-ground material in a maize plant can be used, perhaps for feeding animals, then the ratio rises to 8.9.

Animals are far less efficient at converting and storing energy than plant crops simply because they begin from a plant base and then digest and absorb the energy-rich materials with a rather low efficiency, usually less than 10 percent. So the output:input energy ratios of animal crops might be expected to be an order of magnitude lower than those of plants, and this is indeed the case. Milk production has a ratio of about 0.3 and eggs 0.16. So, for an animal product, considerably more energy is put into the process than is ultimately retrieved in an edible form.

But energy ratios are not necessarily the major concern of the farmer, for he is more involved with economics than energetics. If the use of artificial fertilizers, pesticides and irrigation proves economically worthwhile in terms of additional output of crop, then the energetic balance is relatively immaterial. It is interesting to note, however, that primitive systems of agriculture have much higher output: input energy ratios than the highly mechanized systems of the developed world. The ratios for the hunter/gatherer societies of aboriginal Australians and southern Africans range from 5 to 10; those for the farming/gathering communities of the South Pacific islands are as high as 20; those for wet-rice cultures can be as high as 50. But although these figures make the primitive systems sound efficient, the actual productivity is quite low. It takes approximately 2.6sqkm (1sqmi) of forest to support a bushman, whereas the intensive energy subsidies of modern farming permit 2,000 people to be supported by 2.6sqkm (1sqmi) of the agrobiome.

As fossil fuels become scarcer and therefore more expensive, however, attempts will be made increasingly to reduce the energy subsidy and yet maintain productivity. If, for example, the organic slurry of waste produced by intensive pig and poultry rearing is used as an alternative to artificial fertilizers in maize cultivation, then it is possible to improve the overall output:input ratio from 8.9 to 17.6. The only problem may be the transport costs involved in taking the slurry from its source of production to the field.

Much effort is currently being put into developing the genetic

▼▶ **Animals that have flourished** in the agricultural habitat, but have not necessarily become pests, occur worldwide. Shown here are some examples. (1) Red kangaroo (*Macropus rufus*) from the Australian grasslands. (2) Virginia opossum (*Didelphis virginiana*) from the North American arable areas. (3) Asian white-backed vulture (*Gyps bengalensis*) from Indian arable areas. (4) Harvest mouse (*Micromys minutus*) and (5) Red-legged partridge (*Alectoris rufa*) both from cereal areas of Europe. (6) Paddyfield warbler (*Acrocephalus agricola*) and (7) Chinese pond heron (*Ardeola bacchus*) both from Southeast Asian paddyfields. (8) Cane rat (*Thryonomys gregorianus*) now a pest in Africa.

resources of crops in order to improve productivity, resistance to disease and environmental robustness. Very considerable advances have been made in this respect, but attempts to enhance the productivity of the agriculture of developing nations have not always met with success. Often high-productivity varieties of crops have proved to be extremely demanding in terms of fertilizer requirement, and this can make them expensive to use. In some parts of the world the replacement of low-productivity, pest-resistant strains of crop plants by new varieties which subsequently prove too expensive or too prone to local pests, has led to disaster.

The simplicity of the agricultural ecosystem—most plant crops are grown in single-species stands—can also make them prone to pests and diseases. A fungal or viral disease will spread rapidly between tightly-packed individuals of the same species. One possible ecological solution to this problem is to increase the diversity of the ecosystem by growing several different species together. In the Philippines, for example, coconut palms are grown with bananas, and often it is possible to grow pineapples and Sweet potatoes in the shade of the taller plants. Not only does this reduce the risks of epidemic for any of the species, but it also means that full use is being made of the resources of the environment.

The agrobiome is found in some form within all of the other biomes of the earth. From reindeer farming in the arctic tundra to irrigation agriculture in the desert, from forestry on the high mountains to fisheries in the seas, as man's population has expanded, so has his need to develop the resources of the planet and to channel as much as possible into his own support.

PDM

URBAN BIOMES

The effect of urbanization on the environment... Microclimates... Pollution... Parasites... Buildings as animal habitats... Garbage attracts scavengers... Parks and gardens... Waterways... Exotic species... Sparrows, starlings and pigeons... Rats and mice... Urban foxes...

When people crowd together in towns and cities, they replace natural ecosystems with man-made ecosystems which are organized for man's survival. In these ecosystems man is the dominant species. Urbanization radically alters the wildlife of an area, mainly because plants and animals are physically excluded by the way in which land is used. It also modifies the climate, although it is not this that determines which animals live in towns and cities so much as their capacity to tolerate and exploit man and his activities. Some animals, such as feral pigeons, rats and cockroaches, are able to do this remarkably well, and are found in cities worldwide together with local species, such as the Pied crow of Africa and the House crow of India. Suburbs and other green areas are colonized by opportunists from the surrounding countryside, such as foxes and blackbirds in Britain.

Covering the soil with hard materials, and the artificial production of heat for space heating and industry, turn cities into "heat-islands", the lowest winter temperatures being on average 1–3C° (0.6–1.7F°) higher than in the surrounding countryside. Consequently there is continuous water loss, and although cities receive on average 5–10 percent more precipitation than their rural surroundings, their relative humidity is less, especially in summer, and the soils dry out quickly. The air of a city contains on average ten times the quantity of dust particles, five times the sulfur dioxide, ten times the carbon dioxide and 25 times the carbon monoxide of its rural surroundings, and the soils tend to be more acidic. Sunlight is reduced by 15–20 percent, and cities are on average 5–10 percent more cloudy and may be twice as foggy. The annual mean wind speed is 20–30 percent less, but tall buildings and street layouts cause strong gusts and eddies.

Differences in land use produce distinct microclimates in temperate cities: open spaces, such as Hyde Park in London, England, are more than 1C° (0.6F°) cooler than surrounding built-up areas, and there may be massive accumulations of carbon monoxide in urban chasms during "rush hours." Water from power station cooling towers warms canals and rivers, affecting the distribution of temperature-sensitive aquatic organisms, but increasing overall production, and keeping the canals and rivers free of ice in winter so that birds can feed. Garbage dumps generate heat, creating conditions in which lizards, slowworms, cockroaches and crickets can live. Pigeons cluster on cold days over hot-air outlets, and in England, Pied wagtails roost in flocks in trees in warm city streets.

There is little information as to how city climates affect wildlife, although seasonal plant growth and bird breeding may be advanced in northern cities. Pollution has more obvious effects. Sulfur dioxide and nitrogen dioxide stimulate the growth of black bean aphids because of changes in the food plants. The tissues of urban pigeons and of ground beetles living near heavy

▲ **Opportunistic nester**—a Song thrush (*Turdus philomelos*) rears its brood in an industrial store. Many bird species take advantage of urban features, where natural habitats have been eliminated.

► **Green corridor.** Urban motorways and railroad tracks, if lined by vegetation, are important wildlife habitats, to such an extent that predators, such as this Common kestrel (*Falco tinnunculus*), can glean a living from the fauna of the grass verges.

▼ **Visitor in two ways.** The Gray squirrel (*Sciurus carolinensis*) was not only introduced from North America to Europe, but it can easily be encouraged to visit gardens by putting out food or even, as here, raiding that intended for birds.

traffic contain several times the concentration of lead found in rural animals. In Europe and North America after the industrial revolution, soot pollution together with the disappearance of lichens (which are sensitive to sulfur dioxide) led to evolutionary changes in moths which rest by day on tree trunks, protected from predation by camouflage. A previously rare black form of the Peppered moth spread until it formed 90 percent of populations in and downwind of industrial areas. With the establishment of "smokeless zones," the relative frequency of the black form has declined. In the 1940s and 1950s, the River Thames in London was so badly polluted that it was little better than an open sewer, devoid of all fish but eels, and with scarcely any birds. An efficient antipollution program was started in 1959; by 1983, 104 species of fish, including salmon, had been recorded, and flocks of ducks and wading birds had returned.

Parasites, such as fleas, lice and bedbugs, can be found wherever man lives under crowded conditions. The uncontrolled growth of shanty towns around cities in less developed countries, where housing, waste disposal and water supply are inadequate, has produced so-called "septic fringes" where mosquitoes and other disease vectors flourish. A distinct urban subspecies of the mosquito that transmits yellow fever, *Aedes aegypti*, differing in color and behavior from the rural subspecies, is found in all warmer parts of the world, breeding in water stored or trapped in containers. Ironically, the encouragement of natural vegetation in the "affluent fringe" of cities in South Africa and California has led to outbreaks of murine typhus and other diseases when domestic pets have picked up ticks or fleas that act as vectors.

Buildings provide a variety of living spaces for animals. Ledges on urban "cliffs" are used as roosting or nesting sites by starlings, House sparrows and feral pigeons, and occasionally by kestrels (the species found in London eats sparrows) and Peregrine falcons (which prey on pigeons). Roof spaces and ventilator shafts in city buildings worldwide house roosting or breeding bats and such birds as swifts; mice live in wall spaces; and Norway (or Brown) rats inhabit drains and sewers. In cooler regions, the warmth of buildings shelters crickets, cockroaches, spiders and other animals unable to survive outside. Three species of cockroaches, able to eat anything organic, have been introduced in the last 200 years from the tropics to temperate regions, where they occur in heated buildings.

A major source of food in cities is organic waste. Scavenging kites, vultures and crows are part of the street scene in warmer parts of the world, and until the 18th century, kites and ravens fed in the streets of London. Urban foxes scavenge in Europe, North America and Australia (introduced from Britain) (see pp126–127), coyotes on the outskirts of a number of North American cities, and jackals and hyenas in India and other hot countries. Garbage is exploited by raccoons in North American cities, by opossums in North America and Australia, and occasionally by badgers and Beech martens in European suburbs. A special habitat for scavenging flies is dog feces; in 1970, at least 56,250kg (123,900lb) was deposited each day in New York, and 78,750kg (173,460lb) a day in London.

Within buildings, as far as the human occupants allow, cockroaches, silverfish, ants, flies, beetles, moth caterpillars, mites, rats and mice eat stored food, textiles, and organic debris or refuse, and are themselves eaten by resident predators ranging from spiders to the geckos and snakes of warmer countries. Concentrations of waste in dumps harbor many of the scavengers found in buildings together with a variety of other sorts of flies, and attract the "street-cleaning" birds. Gulls, particularly Herring and Black-headed gulls, feed at dumps in northern cities, as do Marabou storks in Africa and Black vultures in the Americas; scavenging Polar bears make it quite hazardous to visit some Arctic town dumps.

A variety of insects, birds and small mammals feed on the plants growing in urban wasteland; butterflies and bees visit flowers, and birds, such as linnets and greenfinches in Europe and mannikins and waxbills in Africa, eat seeds. In the 1940s and 1950s, insectivorous Black redstarts nested on ruined walls on the bombed sites of London where Rosebay willowherb flourished, as did the Elephant hawk moth, whose caterpillars eat willowherb leaves.

Ornamental plantings, parks and gardens incorporate a range of habitats and feeding opportunities, and since the late

1960s an increasing number of European and North American cities have an element of nature conservation in their planning. Flowering plants are exploited by nectar-feeding insects, and, in tropical cities, by bats and hummingbirds or sunbirds. Bees do so well that beekeepers are able to maintain hives on top of office blocks. Herbivorous insects may get out of hand when many trees of one species are planted together: the lime trees lining many city streets drip with honeydew defecated by millions of aphids, and the hairy caterpillars of Vapourer moths occasionally reach plague densities on plane trees in London squares, as do those of Gypsy moths on North American street trees. Tall trees worldwide provide nesting sites for birds such as crows and magpies (which are town birds in Scandinavia and Britain), and roosting sites for fruit bats in the tropics. Open, grassy areas are feeding grounds for gulls and starlings in temperate cities, and for ibis, such as the hadada of East Africa, in tropical countries.

Suburban gardens, with their extreme variety in plants and structure, are rich habitats, particularly for birds and insects. More species of butterflies occur in West African gardens than in nearby tropical rain forest, and the population density of blackbirds is higher in English gardens, and their breeding success greater, than in woodland. Small ground-feeding doves occur in gardens worldwide, although the Collared dove of Europe is a recent arrival. Originally from India, it began to spread across Europe in 1930, reaching Britain in 1955.

Rivers, canals, reservoirs and ornamental lakes accommodate a range of aquatic life together with many birds, even in the heart of a city. On the lake in St James's Park in central London, coot and moorhen live alongside introduced pelicans; and wild cormorants, herons and ducks visit to feed, while pochard and Tufted duck have bred. Night herons, rails and bitterns breed within New York city limits on wildlife refuges.

There is rarely space for large vertebrates in cities, but Snowy owls occasionally hunt on Toronto airport, elk (called moose in North America) often stray into Moscow and Helsinki, and alligators have been found in swimming pools in suburban Miami, Florida. Deer are maintained in semiwild condition in large parks in some northern cities.

Exotic species sometimes become naturalized from introductions or escapes: Monk parakeets are established in suburban areas of eastern North America and in California, and Ring-necked parakeets in several parts of the USA and in southeastern England. The Common mynah, a relative of the starling native to India, has been introduced to South Africa and Australia where it is almost exclusively a town bird.

◄▲▼ **Urban aliens.** (1) Common raccoon (*Procyon lotor*). (2) House mouse (*Mus musculus*). (3) Norway rat (*Rattus norvegicus*). (4) Herring gull (*Larus argentatus*). (5) Black-headed gull (*Larus ridibundus*). (6) Feral pigeons (*Columba livia*). (7) House sparrow (*Passer domesticus*). (8) Common starling (*Sturnus vulgaris*). (9) Small white butterfly (*Pieris rapae*). (10) Oriental cockroach (*Blatta orientalis*). Flowers are (a) Rosebay willow-herb (*Epilobium angustifolium*). (b) Buddleia (*Buddleia davidii*). (c) Shepherd's purse (*Capsella bursa-pastoris*). (d) Narrow-leaved plantain (*Plantago lanceolata*).

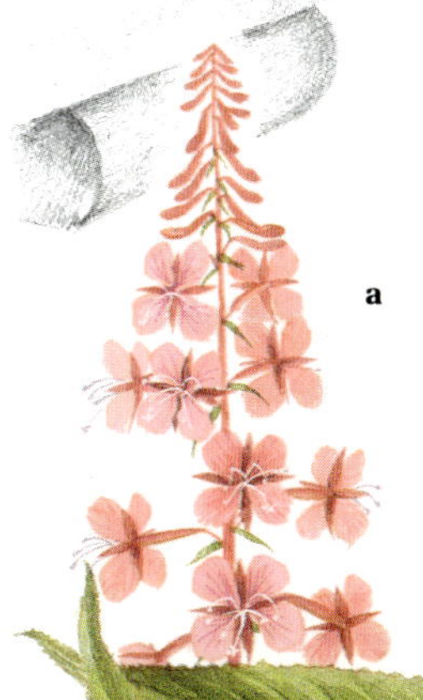

Three European birds which have been introduced worldwide have adapted particularly well to urban life. The House sparrow is predominantly a grain-feeder, and became established in London and New York when horses with their feed-bags thronged the streets. Urban starlings are commuters, roosting communally in city centres, and feeding largely on ground invertebrates in suburbs or the surrounding countryside. Feral pigeons, descended from Rock doves, greatly increased in numbers in cities during the 19th century, and have broadened their diet to include bread and scraps.

Apart from cats and dogs living as scavengers, the only mammals that are truly city-dwellers are rats and mice. The Black rat, native to Southeast Asia, has traveled in ships' cargoes over much of the world, first reaching Europe in Roman times, though it has been replaced in many areas by the larger Norway rat. Norway rats are almost omnivorous, sometimes killing chickens, and do a vast amount of damage, particularly to stored goods, by gnawing at packaging and fouling the contents; they breed rapidly, and are evolving resistance to Warfarin, a poison developed to control them. The worldwide spread of House mice, native to Southeast Asia, is a consequence of their adaptability; they even breed in refrigerated meat stores.

Animals that have adapted to life in buildings or in city centers, and parasites of man, tend to have wide geographical distributions. Which species occur in urban gardens and other green areas, however, depends on the part of the world in which the city is situated. English gardens have robins, Blue tits, and peacock and Small tortoiseshell butterflies; in backyards in eastern North America, there are Blue jays, cardinals, and Black swallowtail and Monarch butterflies; whereas a West African town garden has bulbuls, sunbirds, Citrus swallowtails and richly-colored *Charaxes* butterflies. JO

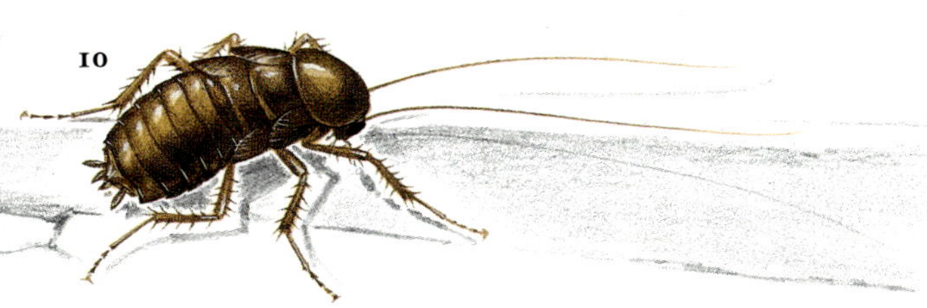

Interloper from the Countryside

The lives of urban foxes

The "concrete jungle" is a cliché which nonetheless encapsulates the widespread despair for wildlife under the onslaught of urbanization. Throughout the world field and forest alike have been devastated by sprawling cities. However, although it does not diminish the tragedy of mutilated wilderness, the urban ecosystem harbors a fauna of its own, made all the more intriguing by its adaptability, a trait nowhere more obvious than in the behavior of city-dwelling foxes.

At 5.5kg (12lb) the Red fox is big enough to be conspicuous, yet in Britain they thrive in many, perhaps most, cities, remaining unseen and undetected by most of their human neighbors. Urban Red foxes are not the only city mammals: in North America, for example, opossums and raccoons live in towns, as of course do countless rodents the world over, together with the feral cats that eat them. Urban foxes are also not unique to Britain: the Red fox is widespread throughout the Northern Hemisphere and occasional sightings have been made in Paris, Stockholm, Amsterdam, Toronto, New York and many other cities. However, it is only in Britain that urban foxes are commonplace, and there they are as urban as you can get: they have been spotted around Trafalgar Square in London, seen sunning themselves on the roofs of terraced houses in Bristol, and traveled along the assembly line of an automobile factory in Oxford.

How do foxes live in an environment which seems so unpromising and which is certainly very different from the nearby farms and woodland? The answer lies partly in the Red fox's phenomenal adaptability, and partly in the fact that there is more wildlife in cities than is apparent at first. A seven-year study of Oxford's city foxes, monitoring their movements by radio tracking, showed that by day city foxes often hide out in the enclaves of wilderness hidden in the town: overgrown orchards, quiet corners of churchyards, railroad embankments and derelict plots. Other fox earths were in more immediately urban surroundings, such as under garden sheds, beneath warehouses or even in the cellars of houses. By night foxes were glimpsed in automobile headlights even in the most built-up districts, and radio tracking revealed that they did indeed occupy every type of urban habitat. Just over a quarter of the part of Oxford where foxes were studied was built up with semi-detached housing, and the foxes foraged in that environment. However, they seemed to prefer detached housing, which only amounted to 7 percent of the area but in which foxes spent 29.9 percent of their active time. Similarly, disused scrubland and woodland amounted to only 4 percent of the town's area, but the foxes singled them out for disproportionate use, 11.9 percent of their time. These foxes spent 21.6 percent of their time in semi-detached housing and 14.0 percent on golf courses and similar parkland. In roaming the city at night the foxes traveled home ranges averaging 86 hectacres (212 acres), and lived in Oxford at a population density of about 2.7 adults/sqkm (6.9/sqmi). Elsewhere, urban foxes can be even more numerous, for example in Bristol there are an estimated 8.5/sqkm (21.8/sqmi).

The diet of Oxford city foxes was surprisingly similar to that of foxes in more rural areas: undigested prey remains in feces revealed that about one-third of the bulk of the diet stemmed from food scavenged from the vicinity of human dwellings—

► **Vulpine head**—portrait of an urban Red fox (*Vulpes vulpes*).

► **Urban diet** BELOW. Diagram showing diet of Red foxes in Oxford, England. The most surprising aspect of the diet of Oxford's foxes, as revealed by a seven-year study, was that almost 30 percent by volume was composed of "wild" vertebrates—wood mice, field and bank voles, hedgehogs, wild rabbits and birds (mainly passerines). These are all prey which the fox might also find in the countryside (although the city foxes ate more rats than normally found in a rural diet). In fact, careful investigation (by extensive live-trapping) revealed that small mammal prey were extremly numerous in overgrown nooks of the town—what the city planner might dismiss as urban wasteland has great potential as an urban nature reserve.

► **Major feeding habitats** of urban foxes BELOW RIGHT. (1) Scavenging from bird tables in competition, but not conflict, with cats. (2) Capturing worms from lawns.

▼ **Night raider**—a Red fox searching for food in Oxford City suburbs. The red collar around its neck holds a radio transmitter which emits signals used by researchers to track its movements.

some feces contained aluminum foil, paper, string, elastic bands and even broken glass. Contrary to popular opinion, foxes only rarely raided trashcans to find this food (dogs are the common real culprit), but instead they found it on bird tables or compost heaps. The second most abundant food type was the earthworm, recognizable by tiny chitinous spikes and lots of soil in the feces. Earthworms surface at night when the weather is suitably mild and humid; it is only when they are on the surface that foxes can catch them, and only then when the vegetation is short enough to enable the fox to bite the worm without ending up with a mouthful of grass; therefore the short grass of playing fields and gardens makes an ideal habitat to hunt for worms. Red foxes eat fruit, and in towns they can feast in orchards on windfall apples, plums and cherries, while in scrubland they can find blackberries.

People often jump to the conclusion that urban foxes are pests, simply because the idea of their living in towns is unfamiliar. In fact, urban foxes cause little nuisance. Some people complain of their barking at night, while others find it hauntingly beautiful; some object to finding a den under their garden shed, others take delight in watching the cubs play. The only serious complaint is about the killing of pets, such as Guinea pigs or rabbits, but this can almost always be prevented by building a secure hutch for the pet. Urban foxes have somehow got a terrible reputation for killing cats, but a survey of 14,000 people in Oxford prompted reports of attacks on only three cats. Similarly of 1,939 fox feces examined over seven years only eight contained traces of cat fur (and these cats could well have been scavenged after accidents with automobiles). There is thus little reason to resent the presence of the urban fox, but great reason to admire its adaptability. DWM

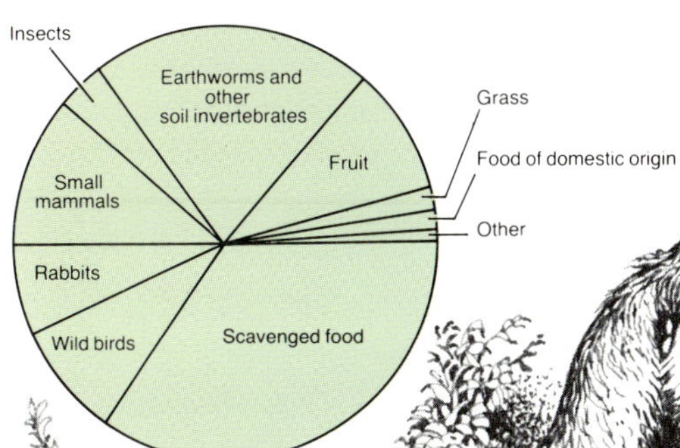

POLLUTION

The nature and extent of pollution by man... Atmospheric pollution and its effects... Pollution of seas and waterways... Pollution from industrial waste... Poisoning from pesticides... Acid rain... The effect of increased carbon dioxide on the climate...

MAN is responsible for releasing vast quantities of different chemical substances into the environment each year, the majority of these being waste products generated by industry and by consumers. For example, every one million inhabitants in a modern city may annually consume over 600,000 tonnes of water, 2,000 tonnes of food and 10,000 tonnes of fossil fuels (coal, oil, natural gas). This leads to the production of 500,000 tonnes of sewage to be discharged into the aquatic environment, 2,000 tonnes of garbage to be incinerated or buried, and about 1,000 tonnes of air pollutants, mostly carbon dioxide and sulfur dioxide. In addition, modern society increasingly relies on a vast range of products composed of many different chemical substances. In total some 80,000 different chemical compounds are marketed; about half of them are produced in quantity, and a great many are released into the environment sooner or later. Many of these are also produced by natural processes, and in some instances the man-made sources are of the same order of magnitude as the natural ones—for example, the atmospheric emissions of sulfur dioxide, oxides of nitrogen and lead compounds. On the other hand, materials such as organochlorine compounds, many pharmaceutical and cosmetic products and plastics are synthetic materials which man has produced for the first time.

The behavior and distribution of man-made substances in the environment depend on many factors, not least of which are the physical and chemical characteristics of the substances themselves and those of the receiving environment. Thus gaseous materials are transported by the atmosphere; dissolved substances become components of the hydrosphere, and particulate matter (depending on its size and density) may be transported by winds and rivers. Persistent substances may remain in circulation for some considerable time. Ultimately they enter the sedimentary environment of lakes, estuaries and the oceans. Many other substances undergo transformation to more innocuous forms or to forms that are potentially more hazardous: thus certain bacteria in aquatic sediments are able to convert inorganic mercury to its more toxic methyl-mercury form. The scale of man's interaction has reached such proportions that the chemical composition of the earth's biogeochemical environment has changed. For example, DDT and polychlorinated biphenyls are present in the fatty tissues of animals throughout the world, the atmospheric concentration of carbon dioxide has increased significantly, and the wet and dry deposits in areas as remote as Greenland and Antarctica contain elevated concentrations of lead and sulfates (derived from sulfur dioxide).

At various stages in the cycling process living organisms are exposed to the substances man releases into the environment. As a result humans and other terrestrial mammals may inhale contaminated air or ingest contaminated food; toxic gases may diffuse into plants; fish may absorb contaminants from the water through their gills; contaminated sediments and soils may adversely affect the microorganisms inhabiting them. Whether or not a particular substance causes damage or harm depends on the level of exposure received and its duration.

Fossil-fuel combustion is the largest single cause of atmospheric pollution. The major primary pollutants are carbon dioxide, oxides of sulfur and nitrogen, and dust particles. In addition, other contaminants (secondary pollutants) are formed in the atmosphere from reactions involving primary pollutants. They include ozone and sulfuric and nitric acids.

The health effects of air pollutants first became apparent when abrupt increases in morbidity and mortality in the general population were associated with acute air-pollution episodes. The most notorious incident occurred in London, England, in early December 1952, when 3,500–4,000 deaths above the norm were recorded during a London smog episode that contained very high concentrations of sulfur dioxide and dust particles. The most vulnerable were the elderly and people with a history of lung and heart diseases. It is generally accepted that the combined effect of sulfur dioxide and particulate matter on the respiratory tract were responsible for the excess deaths and for exacerbating heart and lung diseases in the general population.

Sulfur dioxide is known to affect plant growth adversely, for example, the lichen flora in and around urban/industrial areas of Europe and northeastern North America has disappeared or become impoverished as a consequence of exposure to elevated concentrations of sulfur dioxide.

A different type of smog is produced in many of the world's cities by the action of sunlight on a complex mixture of primary pollutants, mainly oxides of nitrogen and certain hydrocarbons emitted from automobile exhausts. The principal toxicant produced is ozone, but other secondary contaminants include aldehydes and peroxyacyl nitrates. Collectively they produce photochemical smogs which reduce visibility, have an unpleasant smell and are corrosive. Certain susceptible people, particularly heavy smokers, asthmatics and performing athletes, experience difficulty in breathing during periods of photochemical pollution. The worst affected cities are Los Angeles and Tokyo, but many other cities such as Sidney and Athens experience photochemical smog episodes.

Ozone can be transported hundreds of kilometers outside city areas, and is a common air pollutant in the USA and elsewhere. More crop plants and forest trees are injured by ozone than by any other pollutant. In the Los Angeles area alone, continued exposure to elevated ozone concentrations since the

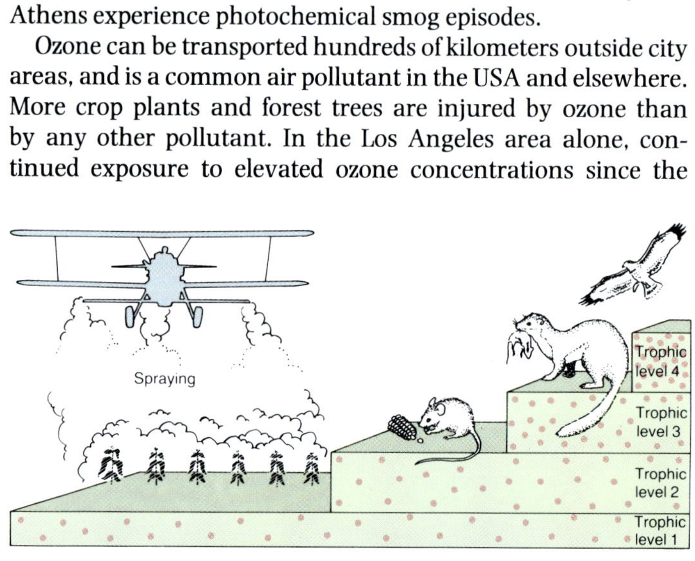

▲ **Invisible killer.** The naked hills around Queenstown, Tasmania were once covered in dense vegetation. Now sulfur fumes from a copper mine have denuded the slopes and stained the rocks hues of chrome, purple, gray and pink.

◄ **Pyramid of destruction.** The fact that toxic chemicals, such as DDT, are not excreted but remain in the cells of the body causes the concentration of this pollutant to increase progressively from link to link in a food chain. Those predators at the top of the chain, whose populations are usually low anyway, become particularly susceptible.

The Earth's Greenhouse

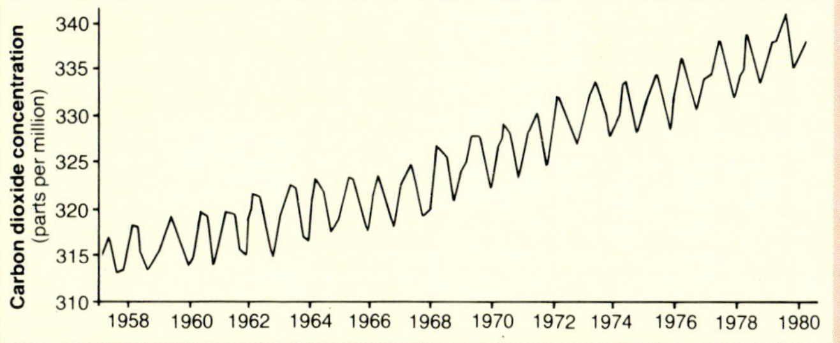

Carbon dioxide in the atmosphere (0.03 percent by volume) is an important factor in maintaining the earth's thermal balance. It is well known that carbon dioxide is fairly transparent to incoming solar radiation, but strongly absorbs infrared radiation emitted by the earth's surface, so that energy that would otherwise be lost to space is directed back to the earth's surface. By this so-called "greenhouse effect" the earth's atmosphere is kept at a higher temperature than it would be without the carbon dioxide.

Data obtained from stations in both Hemispheres demonstrate that carbon dioxide concentrations have increased by 7 percent over the last two decades and probably by 10–15 percent since the beginning of the industrial revolution (see ABOVE).

It is generally agreed that the increased atmospheric burden of carbon dioxide will result in an increase in the temperature of the lower atmosphere and of the earth's surface, but there is considerable uncertainty about the magnitude of the increase. There is also uncertainty about the future rates of increase of carbon dioxide. Recent estimates suggest that concentrations will double by about the year 2080. Scientists predict that this will produce appreciable changes in climate, with an increase of 2–3C° (3.6–5.4F°) in mean global temperatures. In addition there would be notable regional climatic changes, with latitudinal differences in temperature being reduced, and changes in the global pattern of rainfall, with some areas becoming drier (eg the already dry subtropical regions) and others wetter (eg the tropics).

1940s has caused large-scale damage to sensitive pine trees in the San Bernadino Mountains, and reduced yields of citrus fruits, grapes and commercial cut flowers. Ozone concentrations sufficient to cause leaf damage in susceptible vegetation occur in many European countries.

The stratosphere is not free from man's influence. Chlorofluorocarbons (which are used in aerosol sprays and as refrigerants) and other halocarbons (such as trichloroethane) diffuse into this region and interfere with the reactions involved in producing and destroying stratospheric ozone. The net effect is a possible depletion of stratospheric ozone, and in turn an increase in ultraviolet radiation received at the earth's surface. It is predicted that one outcome of these events would be a higher incidence of certain types of skin cancer in the general population.

The disposal of human waste is still one of the major factors affecting the quality of aquatic systems. Discharge of untreated sewage into freshwater systems introduces a variety of fats, proteins and carbohydrates which provide an additional source of food for the organisms in the water. Oxygen is consumed as the organic matter is broken down, and eventually the demand for oxygen exceeds the supply. Extensive areas become anaerobic and consequently uninhabitable for aerobic organisms. Fish are among the most sensitive and the tubificids or blood worms are among the most tolerant of these conditions.

Primary productivity in natural water systems, especially lakes, is frequently controlled by the rate of supply of nitrates and phosphates. Additional inputs of these nutrients into lakes, contained in the runoff from regularly fertilized farmland and treated sewage effluents, accelerate the natural nutrient-enrichment process. During the growing season, large amounts of organic matter are produced as primary productivity is enhanced. Later, more oxygen may be consumed in the breakdown of organic matter than is produced during its synthesis, and so much of the subsequent decomposition occurs by anaerobic processes. Very often these conditions are found in the deeper layers of a lake; for example, it was common to find over 5,000sqkm (2,000sqmi) of the lake bed of Lake Erie deoxygenated during the 1960s. Gross changes in the fauna, notably a decrease in the diversity of species, take place in lakes affected in this manner.

Hydrocarbons are a very diverse group of substances; they are emitted into the atmosphere as gases and particulates from fossil-fuel combustion and by the petrochemical industry. However, they have received most public attention whenever major oil spillages have occurred in the marine environment. For example, the Amoco Cadiz disaster released 223,000 tonnes of crude oil just off the coast of Brittany in France. When such an event occurs, populations of all organisms, coated with newly spilled oil which still contains the toxic light-aromatic compounds, show a sharp rise in mortality. For example, along a 4km (2.5mi) stretch of beach in Brittany which was very badly affected, an estimated 25 million dead invertebrates were washed ashore.

As natural processes degrade the oil, there is a gradual recovery and the former ecosystems become reestablished. On rocky shores this may occur in a matter of weeks, whereas in the bottom sediments and marsh areas full recovery may take a decade. Some animals such as fish and large crustaceans are able to escape or avoid the oil-polluted waters, returning to them when conditions are more favorable.

Populations of sea birds (such as auks, penguins and diving ducks) are the most vulnerable to oil spills. A coating of oil removes the natural oils, leading either to drowning from loss of buoyancy or freezing from loss of thermal insulation. Excessive preening leads to the ingestion of toxic oil droplets. As a result some species in affected areas become severely depleted, and these local losses can have an impact on the entire population of a species.

Many metals have been in everyday use for centuries, and there are numerous examples where, as a consequence of mining, processing or using a particular metal, health problems and ecological damage have resulted. The mining and extraction stages often generate metal-rich waste tips which are acutely toxic to both plants and animals. Runoff from these tips can create toxic conditions in rivers. Very often during the smelting and processing stages, metal-rich fumes are dispersed via the atmosphere to the surrounding countryside, and particulate deposits contaminate soils and vegetation surfaces. There are instances where cattle and horses have died from ingesting vegetation contaminated with, for example, lead and zinc.

In humans, the clinical manifestations of acute lead poisoning have been known for many centuries. People residing near busy roads in inner-city areas may be subject to low, continuous intakes of lead which originates from vehicles using gasoline containing alkyl lead compounds. Children are considered to be the most susceptible, and there is some evidence that the levels of lead they are exposed to are sufficient to impair intelligence and performance, and to be a cause of behavioral disturbances. Thus the use of lead-free gasoline is required by law in some areas. Chronic doses of lead also inhibit several stages of the synthesis of heme (a component of blood).

Of the many different chemical compounds of mercury, the most toxic are the short-chain alkyl- (methyl- or ethyl-) mercury compounds. In humans, the symptoms of methylmercury poisoning reflect damage to the central nervous system, since this compound tends to concentrate in specific areas of the brain, notably the cerebellum which regulates balance, and the visual cortex.

In the general population the largest number of cases of

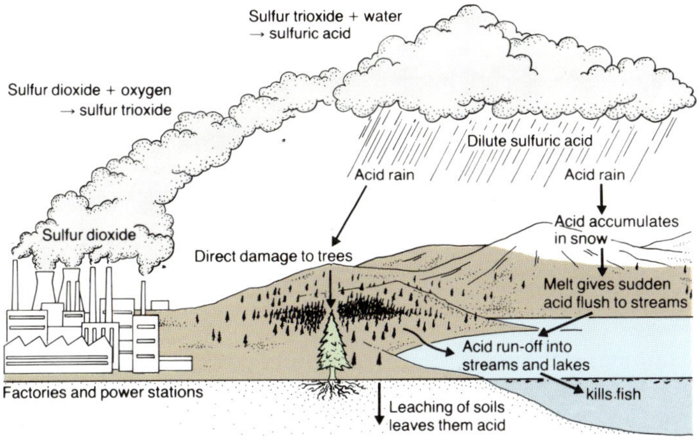

◂ **Black death.** A Common murre or guillemot (*Uria aalge*) killed by oil pollution. A coating of oil can cause drowning or freezing, as well as poisoning through ingestion.

▴ **Acid rain**—a diagram illustrating its origins and effects.

▾ **Last chance**—a Mute swan (*Cygnus olor*) nesting in a polluted lake. Further pollution of this habitat will probably cause the nesting pair to move to another site. Practical habitat management will be required to prevent this happening.

poisoning and of fatalities has been caused by the ingestion of bread prepared from wheat and other cereals treated with alkyl-mercury fungicides. The largest recorded epidemic took place in the winter of 1971–1972 in Iraq when 6,000 people with symptoms of mercury poisoning were admitted to hospital, of whom 500 died. In the Minamata Bay and Niigata areas of Japan, two major epidemics of methyl-mercury poisoning were caused by the industrial release of methyl- and other mercury compounds into the aquatic environment. Nearly 2,000 people suffered neurological disorders and about 400 died through eating fish contaminated with mercury.

Of the many synthetic compounds introduced into the environment, the organochloride pesticides, notably DDT and the very closely related substances DDE and DDD, are the most commonly studied. They are a persistent group of substances which remain in the environment for many years. In living tissues DDT is converted to DDE, and it is this form which is universally present in the fatty tissues of the world's fauna.

Large-scale population declines of many kinds of birds, but mainly fish-eating and bird-eating species, have been ascribed to exposure to organochlorine compounds, in particular DDT. In Europe and North America the most dramatic declines have occurred in Peregrine falcons, sparrowhawks, ospreys and Bald eagles. The majority of bird species also exhibited a significant decrease in the thickness of their eggshells. By the middle 1970s eggshell thinning was demonstrated in at least 54 species of birds belonging to 10 different orders, and in extreme cases resulted in reproductive failure. The main agent causing eggshell thinning is DDT (and DDE), although other organochlorine compounds cause additional harmful effects.

Sulfur dioxide and oxides of nitrogen, emitted from the industrial areas of Europe and North America, are more widely dispersed and remain in the atmosphere for longer periods as a result of the practice in recent decades of releasing pollutants from taller and taller chimneys.

In the atmosphere these compounds undergo various reactions involving oxidation and hydrolysis, to produce aerosols containing dilute concentrations of sulfuric and nitric acids. As a consequence, precipitation over extensive areas of north-eastern North America and northwestern Europe has become increasingly acidic (the mean annual pH is now 4–4.5). Many of these areas are hundreds of kilometers downwind of the original sources. In contrast, the pH of uncontaminated precipitation is usually 5 or just above. A decrease of one unit on the pH scale represents a tenfold increase in acidity.

In the regions affected by acid deposition, numerous fresh-water lakes and rivers have become markedly more acidic, especially those based on geological substrates with a limited capacity to neutralize acid inputs (eg granites and other siliceous rock types). In turn this has led to a rise in the amount of dissolved aluminum in the water. These adverse conditions have caused widespread reductions in biological productivity, and the disappearance of fish populations, notably salmonids. In Sweden 2,500 lakes have suffered fisheries damage and a further 6,500 show signs of acidification; in southern Norway 1,750 lakes lack fish and 900 others are seriously affected; in eastern Canada about 52,000sqkm (20,000sqmi) of surface waters are undergoing acidification. SS

PESTS AND PEST CONTROL

What constitutes a pest?... Sometimes only a nuisance, sometimes positively dangerous... Crop and tree damage from pests... Household pests... Sleeping sickness... Rats and mice... Colorado beetles... Plants which produce chemicals to combat insect pests... Barrier control... Pesticide control, using inorganic, organic and natural substances... Biological control, using natural enemies and sterilization methods...

IN THE natural world there is no such thing as a pest. Animals coexist together in communities linked by a complex set of interrelationships that have evolved through time. These interactions control to a very great degree the population sizes of the individual species in the community.

What then is a pest? A pest is an animal species that interacts with man and causes him some sort of injury. These injuries are of two main types, although the distinction between them is sometimes blurred. Pests may cause actual physical injury as when, for example, a mosquito or a flea bites or a parasitic organism causes a disease. Alternatively, pests cause economic injury by damaging man's livestock, his crops, his stores of food and the things he makes. The term pest is usually applied to animals, but it is also used more generally to cover weeds and fungi as well. A pest is therefore a species that is recognized as having an adverse effect on man's wellbeing. It is this recognition that is particularly important, for without it the interactions of other species with man would be an accepted part of his lifestyle. Such passive acceptance is found in some human cultures.

Population size is sometimes an important element in the pest status of a species. Mosquitoes are not important pests in most of Great Britain, and especially not in towns, but during the summer in the tundra regions of Canada, Greenland and Asia they are extremely abundant. They collect in huge swarms around any animal, including man, that can provide them with a blood meal. Very many species interact with man, but they only rate as pests if their numbers are so high as to cause noticeable injury.

In other instances the pests are primarily a nuisance, not causing any really definable injury. This nuisance status is usually due to the sheer numbers of individuals of the pest involved. Starlings and pigeons roost communally at night and create a considerable mess at the roost site through their nocturnal defecation. Large numbers of birds at airports are, however, hazardous. A number of aircraft crashes during takeoff or landing have been attributed to bird flocks—they can get sucked into the aircraft's engines and stop them functioning.

The type of injury caused is another important consideration. Although in the tundra example mentioned earlier the mosquitoes cause very considerable irritation, they only appear for a short period of time during the year. Mosquitoes in many tropical countries are usually present throughout the year in lower numbers than in the tundra's short arctic summer. However they constitute a far greater threat to man, because of the parasites they can carry, for example *Plasmodium* species which cause malaria, the nematode *Wuchereria bancrofti* which causes elephantiasis, or the yellow-fever virus.

Man gets all his food directly or indirectly from plants. Crop plants worldwide are severely damaged by over 1,000 species of nematode worms and more than 10,000 insect species, not to mention the vertebrate pests such as birds, rats and mice. Overall, man loses about one-third of his crops per year, either while they are growing or in storage. Losses are even greater than this in some countries. In Latin America over 40 percent is lost. Despite its advanced technology and pesticide control, the USA still loses about 30 percent of its crops annually at an estimated value of $20,000,000,000.

Pests impinge on man in other ways. Inedible crops such as timber are grown for the paper and construction industry, and cotton, sisal and jute for cloth and rope making. These too are attacked. The Food and Agriculture Organization has estimated that without insecticide control over 50 percent of the cotton crop grown in the developing countries would be lost. Growing trees are attacked by a variety of insects. Some have a massive effect. The Spruce bud moth occasionally becomes very abundant in the boreal forests of North America and causes extensive mortality among several different tree species, including the Balsam fir and the White spruce.

Several members of the beetle family Anobiidae which normally live in dead and dying timber in forests have become important pests in the house, weakening floors and rafters by the burrowing activities of their larvae. It has been estimated by one of the major British pest-control companies that in southern England about 85 percent of houses built before 1940 have been damaged to some degree by woodworm. In the tropics wooden structures are attacked by termites.

Beside damaging the structure of houses, there is a range of household pests which have a high nuisance value as well as causing injury. Houseflies, greenbottles and bluebottles are all attracted to uncovered food, and in feeding from it may transfer potentially pathogenic bacteria from carrion, fecal

▲ **Garbage raider.** The Norway or Brown rat (*Rattus norvegicus*) is now an international pest of human settlements. Not only does it cause loss to food stuffs and damage to property, but it can also transmit serious diseases to humans.

◄ **"Feathered" locust.** In Africa, the small seed-eating Red-billed quelea (*Quelea quelea*) are considered to be the most destructive of birds. They engage in mass migrations which are correlated to the rains and therefore to available seeds and crops, which they decimate.

material or rotting vegetation on which these flies normally lay their eggs. Cockroaches, similarly, have recently been implicated in the transmission of organisms causing intestinal diseases.

During the winter and early spring, birds frequently come into conflict with the gardener and farmer. Wood pigeons eat substantial quantities of brassicas (cabbages, Brussels sprouts and cauliflowers), and a number of birds take seeds and developing buds.

Clothing and furnishings are not immune from pests. Besides woodworm, which used to be frequently found in cheap softwood used for backing or drawer bottoms and was therefore called the furniture beetle, there are several other beetles of importance. The carpet beetle and the museum beetle are common, feeding on furs, stuffings and leather. The clothes moth attacks clothing, particularly if it is soiled, and may even damage synthetic fibers, although it cannot digest them.

In all the varied activities of man, he either upsets some naturally stable system or creates a new and favorable situation to be exploited. Seals and dolphins have lived in a stable relationship with the fish they feed on for many thousands of years. As man makes increasing demands on the ocean's fish stocks, these mammals are seen as competition—as pests to be

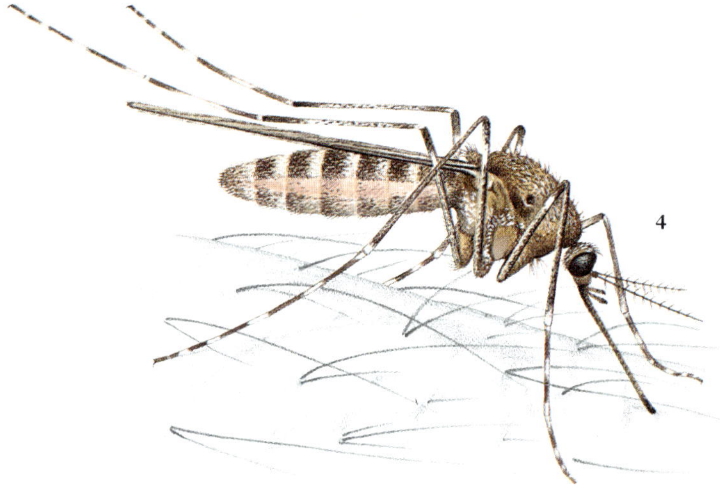

4

destroyed. Large tracts of land in Central Africa were virtually uninhabitable by Europeans because of a disease called sleeping sickness. The causative agent of this disease is the flagellated protozoan, *Trypanosoma*, that invades the blood. It is transmitted principally by the Tsetse fly. People infected by this parasite ultimately become lethargic and sleepy, and usually die. Related trypanosomes attack livestock, especially cattle. Native wild animals act as a reservoir of the trypanosomes but seem to show no ill-effects themselves. The host and the parasite have coevolved together. But in unnatural hosts, domestic cattle and Europeans, the parasite causes very severe symptoms and death. Here man has tried to impose his lifestyle on a natural system, and in the process pest problems have emerged.

Pests most frequently come from the ranks of species already adapted to exploiting new and changing habitats, the so-called *opportunists*. Their two principal characteristics are good powers of dispersal and a rapid rate of reproduction. Size is also important since, in general terms, the larger an animal is the longer it takes to grow to adulthood and the lower its reproductive rate. High rates of reproduction therefore tend to be found in small organisms. Because of their small size and their domination of the animal kingdom in both number and variety, many insects are pests. Among the mammals, the relatively small, fast-reproducing rodents are the main pests.

Since the time of the first great explorers there have been both deliberate and accidental movements of plants and animals from one place to another. Rats and mice, natives of Europe, have been spread to most parts of the world in this way by hitching lifts on ships. Rats, cats and mongooses (ironically, the latter two were used to control the rats) are all implicated as a major cause of the extinction of many formerly endemic species of reptiles, birds and small mammals on many of the Caribbean, Indian Ocean and Pacific islands. These unique island species had evolved in isolation and were unable to cope with the aggressive imports.

Man cultivates only a very small number of the many plants that grow on this planet, but these few are grown in vast tracts of land in all parts of the world where the climate is suitable. This creates many pest problems. The Colorado beetle was a rare beetle of the Rocky Mountains. As the settlers moved west across the Great Plains of North America they brought with them several crop plants, including potatoes. The rarity of the Colorado beetle was probably due to its scarce and scattered food plant, which happens to be related to the potato. Given large fields of potatoes, the beetles rapidly multiplied and spread, causing severe losses to the potato harvests. Travel to and fro across the Atlantic soon brought the Colorado beetle to Europe, where it is now established and—as in the USA—has to be controlled by insecticides. It is not established in Great Britain, and a vigilant watch is kept each summer in Kent for any sign of the beetles which may be blown across the English Channel from France. This example illustrates two of the commonest ways in which pests become established. New crops introduced to an area or country may provide a suitable and abundant food for an animal previously limited by its food. Animals, especially insects, are often controlled to a considerable degree by predators and parasitoids. An animal taken to a new locality will be released from these controlling influences and, provided the climate and food plants are suitable, will be able to multiply rapidly to become a pest.

Pest Control

The basic ecological principle in pest control is very simple. The two main features of pests are that they cause injury and that they tend to have high reproductive rates. Pest control is aimed at reducing the injury to an acceptable level by reducing the numbers and rate of increase of the pest. The objective of control programs is not necessarily total annihilation of a pest;

1

2

3

5

indeed, it is important in some types of control that the pest remains, even though at low density.

There are a wide range of types of control, but they can conveniently be grouped into three main categories. Pests may be prevented from reaching their target by the use of barriers of different sorts; they may be killed directly; or they may be affected by biological control methods.

In tropical countries it is common practice to sleep at night under a mosquito net. This is a physical barrier to the mosquitoes, and not only ensures a good night's sleep without the irritation of their bites but also prevents them passing on any diseases. Crops grown in a greenhouse are to a large extent isolated from plant pests outside, and the use of closed bins and silos prevents access to crops when they are being stored. On a grander scale, the sea provides an effective barrier, as for example in the case of the Colorado beetle, which is common in Europe but cannot easily cross the English Channel to Great Britain. Chemical barriers are created using repellants. Some are reasonably successful, but most are short-lived and many are virtually useless.

Perhaps the most common response to a pest is to reach out for an appropriate pesticide and kill it. This is undoubtedly a

6

7

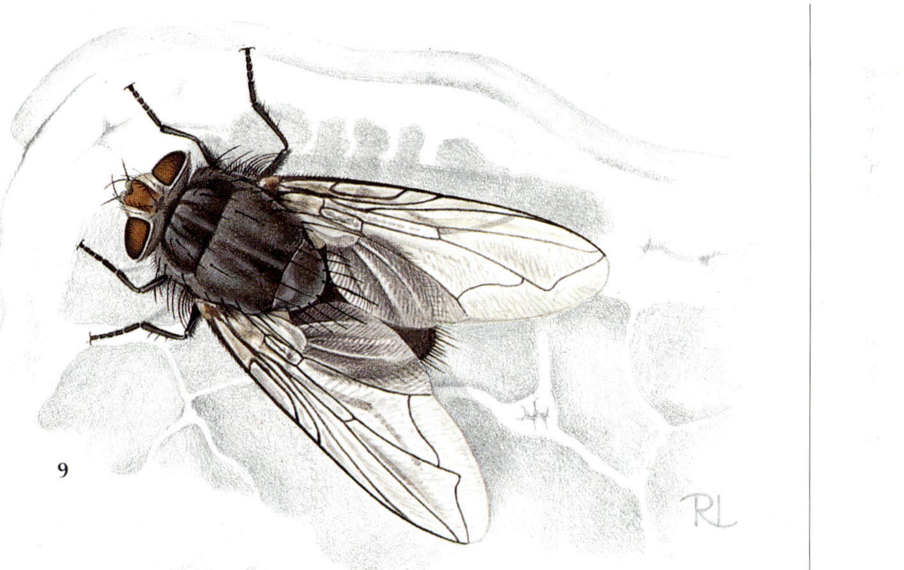

9

8

◄▲ Insect pests of the home and crops. (1) Case-bearing clothes moth and larva (*Tinea pellionella*). (2) Woodworm or Furniture beetle (*Anobium punctatum*). (3) Colorado beetle (*Leptinotarsa decemlineata*). (4) Banded mosquito (*Theobaldia annulata*). (5) Carpet beetle and larva—the woolly bear—(*Anthrenus verbasci*). (6) Bordered white moth and larva (*Bupalus pinaria*), a pest of pine and spruce. (7) Mint leaf beetle (*Chrysolina menthasthri*). (8) Death-watch beetle (*Xestobium rufovillosum*). (9) Bluebottle (*Calliphora vomitoria*).

simple and usually almost instantaneous way of dealing with the pest in the short term. There are ways of killing pests without resorting to chemicals, and such practices are frequently used where food is concerned. Excessive heat or cold are good methods, as is the storage of foodstuffs in atmospheres devoid of oxygen or with high carbon dioxide levels.

Pesticides can be classified into three main groups, inorganic, organic and natural compounds. Some of the earliest pesticides were inorganic. Sulfur, for example, has been burned as a fumigant since biblical times. Arsenic in the form of Paris Green (copper arsenite) or lead arsenate was used up to the 1940s as a stomach poison to control leaf-eating insects such as the Colorado beetle; and phosphorus was used as a rodenticide. These substances, highly toxic to humans, may build up in the environment, causing long-term problems of soil contamination and heavy-metal pollution. They are not used today.

There are four categories of organic compounds used in pest control, and the emphasis has primarily been on the development of new insecticides. The oldest group contains the organochlorides or the chlorinated hydrocarbons, which include such notorious substances as DDT (Dichloro-Diphenyl-Trichloroethane), dieldrin, endrin (also used as a rodenticide), and lindane. What was totally unforeseen was the way in which DDT was stored in the fatty tissues of non-target animals. Through feeding links the DDT became concentrated in the top carnivores, where it caused a number of problems.

To counteract the long-term persistence of these organochlorides a new series of short-lived insecticides was sought. Called the organophosphates, these are related to the nerve gases developed in World War II. Although they are generally more toxic to vertebrates than the organochlorides, they are very short-lived and not persistent. The organophosphates are perhaps the most widely used insecticides today.

Some organophosphates, the so-called systemics, are relatively water-soluble and can be carried in the sap flow of plants. A number of the carbamates, the third main type of organic insecticide, also have systemic properties which makes them more specific to the insects actually feeding on the plant tissues.

The synthetic pyrethroids are the latest developments in insecticide research. These are based on pyrethrum, the naturally-occurring insecticide which comes from the plant of the same name, a chrysanthemum species. Pyrethrum is an attractive insecticide because it has a very low toxicity to vertebrates and an extremely rapid "knockdown" effect on insects. Many insects can recover from the effects of pyrethrum unless it is used together with a slower-acting but more certain insecticide. The synthetic pyrethroids can be used at extremely low doses, and have a greatly enhanced potency.

Besides pyrethrum there are several other natural or botanical insecticides which are becoming more popular with the

► **Man's loss** ABOVE. Nothing but stalks of this Australian pea crop remain after it has been raided by a swarm of locusts.

◄► **Nature's own control of pests.** Natural predators of pests, and those introduced outside their native areas by man, can be an efficient, pollution-free, method of pest control. LEFT Ants (*Anoplolepis longipes*) attack the Cocoa weevil (*Pantorhytes plutus*), a serious pest of cocoa, as here in Papua New Guinea. RIGHT Ladybugs are predators of aphids throughout the world. This species (*Chilomenes lunata*) is seen eating aphids in the Transvaal, South Africa. Ladybugs are often artificially cultured and then released into the wild.

development of the "back to nature" movement. They are generally less effective than synthetic products, but, being natural substances, they break down rapidly and cause no residue problems. Nicotine from the tobacco plant is highly toxic to humans and food crops cannot be harvested for a week after treatment. Another botanical insecticide, rotenone, is commercially extracted from the roots of two tropical genera of legumes (*Derris* and *Lonchocarpus*). It enjoys considerable favor, and its only problem is that it is highly toxic to fish. The Amazonian Indians use it in fishing.

The alternatives to using barriers or pesticides are the biological control methods. Modern insect pest control using predators or parasitoids started in California at the end of the 19th century, with the control of the Cottony cushion scale insect of citrus orchards. A particular species of ladybug was imported from Australia, where it was suspected that the pest had originally come from, and the results were amazing. In a very short time the scale insect was no longer a pest. The remarkable success of this first control led to a growing interest in biological methods during the first 40 years of the 20th century, with some considerable successes, especially in the tropics. With World War II came—among other things—DDT, and interest suddenly switched away from biological control to synthetic pesticides. Now, with the increasing resistance to pesticides, there is renewed interest in natural biological methods, particularly in the field of insect control.

The ecological implications of these two approaches are quite different. Chemical control leads to unstable, oscillating pest populations. The pest population builds up until it is noticed and a pesticide is applied to knock it back. The pest then builds up again and the pesticide is again applied. This results in a "saw-tooth" fluctuation in the pest population, and in the longer term the pest may gradually increase its numbers as resistance develops. What is more, pesticides affect other species. In the case of insects, the use of insecticides frequently affects the slower growing and less common predators and parasitoids far more than the pest species, so that any natural controls that were present are destroyed. This disturbance of the community, by the removal of carnivores, is also a recipe for instability and fluctuation in pest density.

In contrast, biological controls usually add to the complexity of the community with the aim of reducing and stabilizing the population density of the pest permanently. The number of predators and parasitoids may be increased directly by introduction, as in the control of the Cottony cushion scale insect, or more indirectly by changing agricultural practices.

Other biological control methods involve tampering with the pest's reproductive activities. One technique here is to use the insect's own sex attractants (pheromones) to lure males to some sort of trap. Alternatively, the crop may be sprayed with sex attractant so that the males cannot pick out the smells of individual females from the all-pervading background. Reproductive activity is also interfered with using sterilization methods. There are some compounds (antimetabolites and alkylating agents) that cause sterilization, and these can be sprayed into the pest's habitat. They tend to be non-specific, however, and are potentially harmful to humans. Ionizing radiation from a cobalt-60 source has been used to great effect to sterilize male Screw worm flies in culture. These flies are then released to mate with the fertile field population. The eggs of such matings are infertile and die.

With biological control methods there are no problems of environmental pollution. Pesticides do not cause only local contamination. The finding of DDT in the ice and snows of the Antarctic, thousands of kilometers away from the nearest place where it had been used, shows just how much movement is possible with persistent pesticides.

There will, however, always be a place for pesticides in the control of pests. Biological controls are not possible in some situations. In particular, they do not seem to be so successful in dealing with native pests, especially in temperate regions where the weather exerts a strong influence on survival. In such situations using pesticides is, at the moment, the only way to control these pests.

BDT

Pestilential Cycles

The rise and fall of rabbits as pests

The European rabbit originated from the arid areas of northwest Africa and the Iberian peninsula, where it is not noted as a pest. However, successive introductions to other areas have resulted in its becoming a pest of major importance. The rabbit was first introduced to western continental Europe 2,000 years ago and from there to the United Kingdom 1,000 years later. Introductions were also made to Australasia at the beginning of the 19th century, and to Argentina 100 years ago, from where they spread to Chile.

Rabbits feed primarily on grasses, mainly damaging cereal crops and pasture. Cereal fields are invaded from the headlands and rabbit grazing decreases with distance into the field. Plants have to be grazed for several weeks before there is a detectable loss of yield at harvest, so the earlier that grazing begins, the greater the loss. Rabbit grazing also results in a dramatic increase in the abundance and diversity of weed species, due to decreased competition, and delays the ripening of the crop by up to three weeks. This can also cause greater infestations of aphids in summer cereals. In pasture, eight rabbits can eat as much grass as a single sheep, which can result in overgrazing; fine grasses are replaced by coarse ones and by weeds disliked by rabbits. Damage to forestry trees, through removal of bark, occurs in several countries, notably Chile and Argentina, where the young saplings sometimes cannot be established.

The first record of damage in the United Kingdom was in 1340, when rabbits belonging to the Bishop of Chichester escaped and destroyed a field of wheat at West Wittering, Sussex. Damage, however, did not become widespread until the end of the 19th century, reaching a peak between the two World Wars, only to decrease at the advent of the disease myxomatosis. In the United Kingdom, annual yield losses of arable crops and pasture grasses in the early 1950s (before myxomatosis) were estimated to be in the region of £550 million at today's prices. In France, losses in cereals were put at 1 tonne per hectare (0.4 tonne per acre) nationally, compared with 0.76 tonne per hectare (0.3 tonne per acre) in the United Kingdom.

Within 30 to 40 years of their introduction to Australia and New Zealand, rabbits were out of control and could not be contained in either country, even by an ambitious series of fences. In Australia, annual damage to the sheep industry in the late 1940s, before myxomatosis, was put at 0.45kg (1lb) per rabbit for wool with a total value of £270 million lost, and with a total lost lamb production of £110 million, again at current prices. There was a similar total monetary loss in wheat. In New Zealand, where there is no myxomatosis, the value of the pasture in some areas had decreased to such a level by the end of the 19th century that it became more profitable to hunt the rabbits for their skins than to raise sheep; in other areas tens of thousands of hectares had to be abandoned.

There are three facets of European rabbit biology that, in combination, are primarily responsible for its pest status, in contrast to the other 44 species of rabbits and hares. Although the European rabbit is a selective feeder, generally avoiding poisonous, aromatic, downy and very spiny plants, it is considerably more catholic in its food choice than, for example, either the Brown hare or the Mountain hare, in that it will feed on many different growth stages of the same plant. Unlike all other

▲ **Potential pests**—two-week old European rabbits (*Oryctolagus cuniculus*) in their burrow. Litters normally contain five or six young (sometimes up to 12) and they reach breeding condition in three to five months. As many as five litters can be produced every year by each female, which potentially amounts to a lot of rabbits!

▼ **Australian plague.** At times of extreme drought during the Australian plague, rabbits openly congregated at waterholes from the surrounding drought-stricken countryside. The "plague" of rabbits was finally wiped out through the introduction of the disease myxomatosis.

lagomorphs, save the rare Volcano rabbit of Mexico, European rabbits live socially in warrens—a mass of blind and interconnecting tunnels. This results in populations existing at local densities much greater than those of other, solitary species. The European rabbit is highly productive (10 to 30 young per female per year depending on the population density) and has a short life expectancy (less than 1 year). These features, classically, make any species impossible to eradicate through hunting and extremely difficult by other methods.

The myxoma virus of the South American Forest rabbit was first found by Brazilian scientists to be virulent in the European rabbit in 1897. After many attempts, the virus was successfully established in the Murray Valley, Australia, in 1950 with mosquitos, largely *Aedes aegypti*, as the vector. During 1952, the disease was introduced into France and within months had spread over almost the whole of continental Europe. In 1953 the virus was illegally introduced to the United Kingdom, where it is carried by the rabbit flea (*Spilopsyllus cuniculi*). After four years 96 percent of the virus had lost its virulence. This, coupled with resistance to the virus in the remaining rabbits, allowed numbers to increase again; the United Kingdom population is now about 20 percent of the pre-myxomatosis level.

Changes in crop husbandry trends and in the environment following myxomatosis have resulted in less rabbit habitat than before the disease. Rabbits are now less of a widespread problem, but can be important locally. Where they are affecting high-value crops, it is sometimes economic to erect rabbit-proof fencing. This is recommended to be of 31mm (1.2in) mesh and at least 75cm (30in) high, buried at the bottom to a depth of 25cm (10in), with 50cm (20in) turned away from the field towards the invading rabbits. Recently, a portable electric fencing system has become available which could be used to protect a winter-sown crop and then moved to a spring-sown one when the winter crop is no longer vulnerable.

The use of acute poisons to control rabbits was never successful in the past, owing to the rabbits' wariness of the bait and the very high kill (over 90 percent) necessary to achieve control. The fumigation of warrens with cyanogenic compounds is locally effective, but cannot usually be done on a large enough scale to give lasting control. In lowland New Zealand, the synthesized natural product sodium monofluor-acetate has been widely used with some success since 1953. Extensive aerial applications are made once or twice a year and supplemented by shooting and "patch-poisoning;" however, there has been less success with this chemical in the more inaccessible highlands. A new generation of anticoagulant poisons are currently being evaluated for rabbit control in New Zealand, as well as for Chilean forestry. RAB

CONSERVATION

What is conservation?... Understanding ecology... The Western world versus the developing countries... Nature reserves and conservation groups... The destruction of habitats... International action... Major threats to habitats...

It is conservation more than any other area that has brought ecology into the arena of public debate. People are now deeply concerned that their natural heritage should be preserved for future generations.

However, conservation is rather more than preservation. It reflects an attitude to nature and land-use which implies continuity of supply but also caring utilization by man. There are almost as many definitions as practitioners, and they often distinguish between resource conservation (dealing with soils, water or forests) and nature conservation (dealing with native plants and animals). Another distinction is between species conservation and habitat conservation. Concerned people often start by thinking about species such as whales, the Giant panda or butterflies, but in order to protect viable populations of such animals extensive areas of appropriate natural or seminatural habitat need to be managed in a particular way. Thus habitat conservation, in dealing with forest, grasslands or wetlands, is often a more important approach, and if land is managed so that the larger and more conspicuous animals survive then it is probable that other organisms will also have a safe future. This philosophy reflects the fact that the animals which are initially concerned often exist at the top of food chains, and for every large predator such as a bird of prey, thousands of other small animals and even larger quantities of vegetation are needed for support.

In order to protect and manage the plants and animals that are valued and the habitats within which they occur, there needs to be an understanding of the relationships between all the species and the environmental factors that control them. This is the ecological basis for conservation. It involves a knowledge of the flow of energy through ecosystems, such as in the food chains referred to above, and the cycling of nutrients and their conservation as in forest and moorland ecosystems. There are also general principles that the ecologist can communicate to the conservationist about the relationship between plant density and yield. Also there is the interaction between individual plants and between plants of different species which is sometimes referred to as competition. The capacity of species to reproduce is also important, and can vary from species to species; some produce many small eggs, seeds or young and can increase their populations rapidly, others produce few large seeds or young and build up their populations only very slowly. The former include many pests and weeds with a boom-or-bust strategy, whereas the latter are often the concern of conservationists because they need large areas of stable vegetation and replenish their depleted populations slowly, eg tigers and whales. Stress in vegetation is considered to be important, as caused by nutrient deficiency or shallow soils, for example, and this can reduce the dominance of aggressive species, so allowing large numbers of plant species to coexist. This high species-richness or diversity is usually considered desirable, and is characteristic of some successional stages such as temperate chalk grassland and stages in sand-dune vegetation. Ecological science is therefore essential for the rational management of vegetation and the dependent animal populations.

Concern with nature conservation is important both in the Western world and in developing countries. In the former, pressures of increasingly intensive agriculture and forestry as well as pollution from waste and excessive use of fertilizers and pesticides are the chief concerns, whereas human population growth is not. In the developing world, on the other hand, human population growth is still making increasing demands on resources, space and of course wildlife. In these areas pollution is not usually the major problem, although there are some serious exceptions to this generalization. Major threats come from the local needs for food and fuel, especially the production of charcoal for cooking. Tropical forests are being lost at an alarming pace, and the edges of deserts are also deteriorating, a process referred to as desertification. While the human populations of the Western world consume vast quantities of resources (minerals, oil, food and fiber) at about forty times the rate per head of developing countries, the rising population of the latter causes concern. It is hoped that the people of these countries will have a higher standard of living, but the prospect of everyone in the world enjoying the lifestyle of, say, wealthy Americans would be disastrous for the world's wildlife, unless accompanied by strong national and international planning and legislation. So far the only protective legislation appears to have come about at the same time as the crash of particular populations, such as whales, or the virtual loss of particular ecosystems, chalk grassland in Britain, for example.

In Western countries systems have been devised for protecting species and habitats in reserves. These occur as nature reserves, protected areas or within National Parks. Such areas often only represent about 1 percent of a country's area, although a much larger proportion would be desirable. The remainder, for example, in western Europe, North America and Japan, is often under intensive agriculture and forestry. This is the matrix, and it remains the home of most plants and animals. However, it is in this area that change is most rapid, and many plant and animal species are decreasing at an alarming pace. This can only be controlled by the introduction of planning systems to regulate the established land-uses, or legislation to prevent landowners and occupiers from carrying out certain activities. Unfortunately both planning and legislation can be interpreted as negative measures, and it would be better also to think in terms of more positive action such as education.

Governmental agencies for nature protection are usually responsible for designating reserves, although in many Western countries it is the voluntary movements or non-governmental organizations which are also very active. In the USA the Nature Conservancy Council, Ducks Unlimited and the Sierra Club are conspicuous, and in Britain it is the County Trusts for Nature Conservation and the Royal Society for

► **Overgrazed and turning to desert**—domestic goats (*Capra aegagrus hircus*) grazing on the receding vegetation on the edge of the Kalahari Desert. The loss of vegetation has been caused by a combination of overgrazing by domesticated animals and drought conditions.

Nature Conservation that are most active. The latter acts as an umbrella organization for the smaller regional bodies. Some voluntary bodies are very large; the Royal Society for the Protection of Birds has as many as 360,000 members and 80 reserves totaling 36,360 hectares (90,000 acres). At this stage it is worth considering the criteria used by governmental and other agencies to select their nature reserves. The 10 recommended criteria are size, diversity, rarity, naturalness, typicalness, fragility, recorded history, position in an ecological unit, potential value and intrinsic appeal. In practice it may be opportunity and threat which really decide whether areas are acquired or not. In some countries, such as the Netherlands, a system of grants to subsidize reduced levels of agriculture such as not plowing or improving some types of permanent pasture, appears to work reasonably well.

Conservation problems are not evenly distributed across the world's surface, but are concentrated in some areas much more than others. For example, there is particular concern over the rapid exploitation of tropical rain forest, which is almost certainly the most species-rich of all ecosystems. Concern is also expressed about the over-grazing of semiarid and arid areas as well as the heavy pressures of grazing and burning on Mediterranean ecosystems. In the oceans, over-fishing causes concern on a worldwide basis, as do new fisheries such as the exploitation of krill in the Antarctic Ocean. The movement of airborne pollutants, and especially sulfur dioxide in solution, referred to as acid rain, causes concern in Scandinavia, upland areas of Germany and Czechoslovakia, northeastern North America, and parts of Scotland.

On the other hand, the conservation problems of the more technologically advanced Western countries reflect the temperate climate that predominates and the types of cropping that consequently occur. These include the loss of habitats which reflect past agricultural systems such as small fields, permanent

Major Threats to Habitats

(1) Browsing and overgrazing. (2) Clearing of vegetation for agricultural crops. (3) Changes in agricultural methods. (4) Logging and exploitation of forests. (5) Changes in forestry practice. (6) Dam construction, hydroelectric power schemes and flooding. (7) Drainage. (8) Water pollution. (9) Industrialization and urbanization. (10) Mining and quarrying. (11) Road building. (12) Coastal pressures and developments, eg for tourism. (13) Disturbance through trampling and motor vehicles. (14) Fire. (15) Introduced species. (16) Collecting specimens, eg for horticulture, and picking wild flowers. (17) Critically small populations.

► **Raging inferno** in north Queensland, Australia. A huge pasture fire burning after tropical rain forest cover has been cleared. Here, as elsewhere, the wildlife is protected by law—but not habitats.

▼ **Conservation in action**—White rhinoceros (*Ceratotherium simum*) in holding pens. In parts of South Africa conservation has been so successful that populations are too high. Excess animals are used to stock other areas or, more debatably, shot legally by game hunters.

pasture, hedges and wet meadows, and their replacement by intensive arable farming encouraged by subsidies from national or supranational governments such as the European Economic Community. In many countries, deciduous woodland is being lost and replaced by faster-growing conifers, and many of these areas have had a continuous woodland history ("ancient" woodlands) with an assemblage of species not found in secondary woods. The loss of the countryside, as this attrition is called, is currently the subject of a campaign by the "Friends of the Earth" movement, which is active on both sides of the Atlantic. Other major concerns are the loss of wetlands, principally by land drainage for agriculture, which is a worldwide problem, and the loss of calcareous grassland as on chalk and limestone. As mentioned above, these areas are very rich in species, especially flowering plants and butterflies.

International agencies such as the International Union for the Conservation of Nature and Natural Resources (IUCN), the United Nations Environment Program (UNEP) and the World Wildlife Fund (WWF) have been so concerned about conservation problems that in 1980 they jointly launched the World Conservation Strategy with the declared aims of combining conservation with development, preserving the world's genetic diversity, maintaining the world's life-support systems, and sustaining the utilization of species and ecosystems. Now each member country is expected to produce a national conservation strategy, and some have already done so. The UK program consisted of seven review groups whose reports have now appeared in bound form. The exercise also involved the commissioning of an opinion poll involving 2,000 people to assess their concern for their environment. It is significant that 50 percent thought that there was a serious risk of using up the world's resources, and over 40 percent thought that pollution was a major world problem. This brings us back to the beginning of this discussion, that conservation arises from concern by people for resources. Ultimately it is a human-oriented concern, to protect species and habitats for enjoyment. Even if few people have actually seen them, the majority want to know that Giant pandas are crunching bamboo stems, that tropical reefs are not decaying from marine pollution and that sulfuric acid is not dropping from the sky killing trout and salmon in rivers and lakes. However, this demands some sacrifices including at least some restraint on growth, resource utilization, and affluence. It involves further study of the interactions of ecology, agriculture, forestry, economics, sociology, planning and law.

This study, and the furtherance of ecological knowledge, helps fuel concern for conservation. BG

Bibliography

The following list of titles indicates key reference works used in the preparation of this volume and those recommended for further reading. The list is divided into categories corresponding to those of the volume.

General Ecology

Berry, R.J. (1977) *Inheritance and Natural History*, Collins, London.
Blondel, J. (1979) *Biogéographie et écologie*, Masson, Paris.
Colinvaux, P.A. (1973) *Introduction to Ecology*, John Wiley, New York.
Dajoz, R. (1977) *Introduction to Ecology*, Hodder and Stoughton Educational, London.
Krebs, C.J. (1985) *Ecology* (3rd edition), Harper and Row, New York.
May, R.M. (ed) (1976) *Theoretical Ecology*, Blackwell, Oxford.
Putman, R.J. and Wratten, S.D. (1984) *Principles of Ecology*, Croom Helm, London.
Whittaker, R.H. (1975) *Communities and Ecosystems*, Collier Macmillan, London.

Biogeography and Biomes

Ayensu, E.S. (ed) (1980) *Jungles* Jonathan Cape, London.
Bramwell, D. (ed) (1979) *Plants and Islands*, Academic Press, London.
Carlquist, S. (1974) *Island Biology*, Columbia, New York.
Chapman, V.J. (ed) (1977) *Wet Coastal Ecosystems (Ecosystems of the World*, vol 1), Elsevier, New York.
Chernov, Y.I. (1985) *The Living Tundra*, Cambridge University Press, Cambridge.
Cloudsley-Thompson, J.L. (1975) *Terrestrial Environments*, Croom Helm, London.
Cloudsley-Thompson, J.L. (1984) *Sahara Desert*, Pergamon Press, Oxford.
Collinson, A.S. (1977) *Introduction to World Vegetation*, George Allen and Unwin, London.
Cox, C.B. and Moore, P.D. (1985) *Biogeography: An Ecological and Evolutionary Approach* (4th edition), Blackwell Scientific Publications, Oxford.
Goldman, C.R. and Horne, A.J. (1984) *Limnology*, McGraw-Hill, Tokyo.
Hardy, A. (1956) *The Open Sea: The World of Plankton*, Collins, London.
Heywood, V.H. (ed) (1985) *Flowering Plants of the World*, Croom Helm, London and Prentice Hall, New York.
Hora, B. (ed) (1981) *Oxford Encyclopedia of Trees of the World*, Oxford University Press, Oxford.
Larsen, J.A. (1980) *The Boreal Ecosystem*, Academic Press, New York.
Lewis, J.R. (1964) *The Ecology of Rocky Shores*, English Universities Press, London.
Louw, G. and Seely, M. (1982) *Ecology of Desert Organisms*, Longman, London.
Macan, T.T. and Worthington, E.B. (1974) *Life in Lakes and Rivers*, Collins, London.
Moore, D.M. (ed) (1982) *Green Planet: The Story of Plant Life on Earth*, Cambridge University Press, Cambridge.
Moore, P.D. (ed) (1985) *European Mires*, Academic Press, London.
Moore, P.D. and Bellamy, D.J. (1974) *Peatlands*, Elek, London.
Myers, N. (1984) *The Primary Source: Tropical Forests and Our Future*, Norton, New York.
Pearsall, W.H. (1950) *Mountains and Moorlands*, Collins, London.
Pielou, E.C. (1979) *Biogeography*, John Wiley, New York.
Polunin, O. and Walters, M. (1985) *A Guide to the Vegetation of Britain and Europe*, Oxford University Press, Oxford.
Pruit, W.O. (1978) *Boreal Ecology*, Edward Arnold, London.
Ranwell, D.S. (1972) *Ecology of Salt Marshes and Sand Dunes*, Chapman and Hall, London.
Richards, P.W. (1952) *The Tropical Rain Forest*, Cambridge University Press, Cambridge.
Scott, P. (ed) (1974) *The World Atlas of Birds*, Mitchell Beazley, London.
Sutton, S.L., Whitmore, T.C. and Chadwick, A.C. (1983) *Tropical Rain Forest: Ecology and Management*, Blackwell Scientific Publications, Oxford.
Tivy, J. (1982) *Biogeography*, Longman, London.
Walter, H. (1979) *Vegetation of the Earth*, Springer-Verlag, New York.
West, R.G. (1977) *Pleistocene Geology and Biology* (2nd edition), Longman, London.

Man and Nature

Ayensu, E.S., Heywood, V.H., Lucas, G.L. and DeFilipps, R.A. (1984) *Our Green and Living World: The Wisdom to Save It*, Cambridge University Press, Cambridge.
Ehrlich, P.R., Ehrlich, A.H. and Holdren, J.P. (1977) *Ecoscience: Population, Resources, Environment*, W.H. Freeman, San Francisco.
Fitter, R.S.R. (1945) *London's Natural History*, Collins, London.
Harris, D.R. (1980) *Human Ecology in Savanna Environments*, Academic Press, London.
Hoskins, W.G. (1977) *The Making of the English Landscape*, Book Club Associates, London.
Mellanby, K. (1967) *Pesticides and Pollution*, Collins, London.
Mellanby, K. (1981) *Farming and Wildlife*, Collins, London.
Pollard, E., Hooper, M.D. and Moore, N.W. (1974) *Hedges*, Collins, London.
Ramade, F. (1984) *Ecology of Natural Resources*, John Wiley, Chichester.
Ratcliffe, D. (ed) (1977) *A Nature Conservation Review* (2 vols), Cambridge University Press, Cambridge.
Simmons, I.G. (1974) *The Ecology of Natural Resources*, Edward Arnold, London.
Simmons, I.G. (1979) *Biogeography: Natural and Cultural*, Edward Arnold, London.
Simon, J.L. and Kahn, H. (eds) (1984) *The Resourceful Earth*, Basil Blackwell, Oxford.

Picture Acknowledgements

Key: *t* top, *b* bottom, *c* center, *l* left, *r* right.
Abbreviations: A Ardea. AN Agence Nature. ANT Australasian Nature Transparencies. BCL Bruce Coleman Ltd. FL Frank Lane Agency. NHPA Natural History Photographic Agency. NSP Natural Science Photos. OSF Oxford Scientific Films. P Premaphotos Wildlife/K. Preston-Mafham. PEP Planet Earth Pictures/Seaphot. SAL Survival Anglia Ltd. SPL Science Photo Library.

1 P. 2–3 OSF/G. Bernard. 5 Tony Morrison. 6–7 Auscape International/J.P. Ferrero. 10–11 Leonard Lee Rue III. 13 Tony Morrison. 15 A. Bannister. 16–17 FL/C. Carvalho. 18, 19 P. 20–21 PEP/C. Pétron. 24 ANT/R.&D. Keller. 25 NSP/M. Stanley Price. 31*b* BCL/R. Wilmshurst. 31*t* Roger Hosking. 32–33 Peter Veit. 35 Andrew Laurie. 39 George Frame. 42 P. 43 Michael Fogden. 44 J. MacKinnon. 45 Auscape International/G. Threlfo. 46–47 P. 51 A/F. Gohier. 52*t* David Hosking. 52*b* P. 53 R.O. Peterson. 54 AN/Lanceau. 55 William Ervin, Natural Imagery. 56*t* Len Rue Jr. 56*b* A/S. Roberts. 56–57 FL/M. Newman. 58–59 Aquila. 59*r* Len Rue Jr. 62–63 ANT/T. & P. Gardner. 64 NSP/Dick Brown. 65*b* P. 65*t* SAL/J.&D. Bartlett. 66–67 C.A. Henley. 68 ANT/Bruce Thomson. 69 NHPA/S. Dalton. 70–71 OSF/Kathy Tyrrell. 71*b* SAL/Alan Root. 72*t* PEP/Richard Matthews. 72*b* P. 73 Michael Fogden. 74–75 SAL/Jeff Foott. 76 Tony Morrison. 77*t* Leonard Lee Rue III. 77*b* Michael Fogden. 78–79 SAL/Jeff Foott. 80–81 Fred Bruemmer. 81*t* PEP/M. Oglivie. 82–83 SAL/Joel Bennett. 82*b* Eric and David Hosking. 84 SAL/D.&M. Plage. 86–87 Leonard Lee Rue III. 91 Biofotos/ Heather Angel. 94–95 SAL/Alan Root. 95*t* SAL/L.&T. Bomford. 99 SPL/Martin Dohrn. 100–101 S. Bannister. 101*b* ANT/Keith Davey. 104–105 PEP/Warren Williams. 106 Biofotos/J. Hoogesteger. 107*t* OSF/F. Ehrenström. 107*b* B. Picton. 108–109 PEP/Bill Wood. 112–113 A/F. Gohier. 114*b* ANT/M.F. Soper. 114–115 Biofotos/Heather Angel. 116–117 Tony Morrison. 118–119 P. 119*b* A. Bannister. 122*b* A/I. Beames. 122–123 NHPA/M. Leach. 122*t* FL/A.R. Hamblin. 126 M.T. Newdick. 127 D.W. Macdonald. 129 ANT/J.R. Brownlie. 130 David Hosking. 131 NHPA/M. Leach. 132–133 OSF/P. Ward. 133 NHPA/S. Dalton. 136*l* PEP/C. Prior. 137*b* A. Bannister. 137*t* ANT. 138–139 OSF/G. Bernard. 138*b* The Australian Information Service, London. 141 A. Bannister. 142*b* Swift Picture Library/N. Dennis. 142–143 ANT/R.&D. Keller.

Artwork

Abbreviations: OI Oxford Illustrators Limited. DO Denys Ovenden.

4,7 OI. 9 DO. 12,14 OI. 22,23 Roger Gorringe. 26,27,28,29 DO. 30,35 OI. 36,37 DO. 38*t* Wayne Ford. 38*b* OI. 40,41 SD and Wayne Ford. 46 OI. 48,49,60,61 DO. 69 OI 78,79,84,85 DO. 88,89 Mick Loates. 92,93,96,97 DO. 102,103 Roger Gorringe. 111 Mick Loates. 112 OI. 120,121,124,125 DO. 127*l* Wayne Ford. 127*r* Priscilla Barrett. 128 OI. 129 Wayne Ford. 131 OI. 134,135 Richard Lewington.

GLOSSARY

Acid rain precipitation, both rain and snow, made acidic in reaction through the chemical POLLUTION of air by waste gases, such as oxides of sulfur and nitrogen, from industry and car exhausts.

Adaptability the ability of an organism to alter its mode of behavior, or even its physiology, when placed under a new type of stress.

Adaptation the process whereby, under the influence of NATURAL SELECTION, an organism gradually changes genetically in such a way that it becomes better fitted to cope with its environment.

Association a group of organisms which is usually found together, either because they are dependant on one another or because they have similar environmental requirements.

Atmosphere the gaseous envelope that surrounds the earth.

Bergmann's rule when a bird or mammal species is distributed over a range of climate, races from colder regions tend to be larger than those from warmer regions.

Biological control the use of one species of organism (usually a PREDATOR or PARASITE) to control the population of another species.

Biomass the total quantity of organic matter associated with the living organisms of a given area at a particular time.

Biome a major global ecological unit, or type of flora and fauna formation (eg savanna grassland, boreal forest).

Biosphere that part of the Earth which is capable of supporting life; includes part of the ATMOSPHERE, LITHOSPHERE and HYDROSPHERE.

Body size the bulk of the individual, expressed in terms of volume or weight.

Browsing the activity of HERBIVORES which consume the leaves or branches of shrubs and trees.

Carnivore an animal which eats the flesh of another animal, usually involving the death of the latter.

Climate the general summary of weather conditions at a site averaged over a period of time.

Climax stage in development of an ECOSYSTEM when there is no further net growth in BIOMASS; relative to other stages, climax flora and fauna are rich and their interrelationships complex.

Climax dominant species of plant which are the most influential (and often the most bulky) in stable vegetation.

Coevolution the process whereby two associated species of organism evolve together and become adapted to one another, eg flowers and their pollinators, plants and their animal consumers.

Colonization the process of INVASION of a new habitat by plants and animals.

Commensalism an association between two species which require the same food and which may benefit either species and is harmful to neither.

Community a collection of populations of a number of species interacting together.

Competition the interaction between two or more species, or between individuals of a single species, in which a required resource is in limited supply and consequently one or both of the competitors suffer in their growth or survival.

Competitive exclusion when two organisms exploit the resources of the environment in the same way, one (the less fit) will be eliminated by the other.

Conservation the rational management of and care for the BIOSPHERE, in order to avoid the creation of imbalance resulting in the destruction of habitats and the extinction of species.

Continental drift see PLATE TECTONICS.

Cultivation the management of an ECOSYSTEM with the specific intention of channeling energy into human beings.

Decomposer an organism which relies upon the dead tissues of other organisms as an energy source; in using this energy it liberates nutrients from those tissues into the ENVIRONMENT.

Defense strategy the collection of adaptations, whether anatomical, physiological or behavioral, which provide an organism with the ability to withstand the attentions of predators or parasites.

Detritivore an animal which feeds upon the dead remains of other organisms, both plant and animal.

Disease the effect on a plant or animal of harboring pathogenic, harmful microorganisms.

Dispersion the pattern of spatial distribution exhibited by an organism.

Distribution the extent of the geographical range of an organism.

Diversity a measurement of the richness of species in a given area, sometimes also incorporating the evenness of COMMUNITY, ie the degree to which certain species dominate the community in numerical terms.

Ecological equivalent species which occupy the same ECOLOGIAL NICHE in separate areas, often on different continents; they are often unrelated taxonomically.

Ecological niche the summation of the ecological requirements and the role of a species in a COMMUNITY.

Ecology the study of organisms in relation to their physical and living environment.

Ecosystem a unit which includes all the living organisms and the non-living material within a defined area, the size of which is relatively arbitrary.

Endangered species a species whose population has dropped to such low levels that its continued survival is insecure.

Endemic a species which is limited in distribution to a defined area where it is native.

Energy the ability to do work. It may take many forms, such as light, heat and chemical; it changes form as it flows through the ECOSYSTEM.

Environment the surroundings of an ORGANISM, including both the non-living world and the other organisms inhabiting the area.

Environmental stress an environmental factor, such as temperature, water availability or acidity, which attains a level close to the TOLERANCE LIMIT of a species and hence causes problems for its continued survival.

Eutrophication the enrichment of a HABITAT (often acquatic) with nutrient elements such as nitrogen or phosphorus.

Extinction the complete elimination of a population; often used in a global sense, but can be used of local populations.

Foodweb the complex feeding interactions between species in a community.

Fundamental niche the complete range of conditions within which a species is able to survive, seldom realized in practice because of interactions with other organisms.

Grazing the action of a HERBIVORE which feeds upon herbaceous vegetation close to the ground.

Greenhouse effect the accumulation of gases, such as carbon dioxide, in the atmosphere which prevent infra-red radiation leaving the Earth and hence cause increasing global temperature.

Habitat the locality within which a particular organism is found, usually including some description of its character.

Herbivore an animal which feeds exclusively on living vegetable matter.

Hibernation the passing of unfavorable winter in a state of reduced metabolism, when all body processes, including respiration and excretion, are reduced to a minimum.

Hydrosphere the global water mass, including atmospheric, surface and subsurface waters.

Individual a single organism; sometimes difficult to distinguish in colonial animals and in plants.

Invasion the entry of a species into an area which it formerly did not occupy.

Isolation a process which leaves an individual or a population remote from other members of the species.

Life-form the physical architecture of vegetation, defined on the basis of where the perennating organs, eg buds, bulbs, rhizomes, are held.

Life style the mode of life of an organism, including its behavior, feeding requirements and habitat.

Lithosphere the outer part of the earth composed of rock.

Locomotion the means by which an animal achieves movement.

Management of resources the informed manipulation of an ECOSYSTEM by man in order to gain some product from the ecosystem without causing its disruption.

Maturation gaining in age, often referring to reproductive maturity.

Metabolism the biochemical processes of life within the cell and the organism.

Microbe, microorganism organisms of microscopic or ultramicroscopic size, such as bacteria, some fungi, viruses.

Microclimate the climate of a small, defined area, such as a valley or even a hollow log.

Migration the systematic movement of a population from one area to another.

Mobility the ability of an organism to move from one place to another.

Morphology the form and arrangement of organs in the body of an organism.

Natural selection the process by which those organisms which are not well fitted to their environment are eliminated by predation, parasitism, competition, etc, and those which are well fitted survive to breed and pass on their genes to subsequent generations.

Niche see ECOLOGICAL NICHE, FUNDAMENTAL NICHE.

Nutrient cycle the movement of elements around an ECOSYSTEM between living and non-living components.

Nutrient reservoir a component of an ECOSYSTEM (such as the trees, or the soil's organic matter) which acts as a means of storage of a given element.

Ocean currents the major global movements of waters within the seas.

Omnivore an animal which is prepared to consume both plant and animal material in its diet.

Opportunist an organism which as a result of its capacity to migrate rapidly, to breed fast, or to survive in a dormant state, is able to take advantage of an opportunity, such as ECOSYSTEM disturbance, to expand its population.

Organism a living creature, animal, plant or MICROBE.

Parasite an organism which is totally dependant upon another organism (the host) for its energy, is usually very closely associated with its host and which often causes its reduced growth or reproduction but only rarely kills it.

Pest an organism which causes problems for human beings by interfering with their management of ECOSYSTEMS; can include plants, animals and microbes.

Pest control the reduction of pest populations by various means, including chemical and BIOLOGICAL CONTROL.

Plate tectonics/continental drift the slow movement of parts of the Earth's crust due to convectional flow in its liquid core.

Pollution the disruption of a natural ECOSYSTEM as a result of human contamination.

Population a collection of individuals of the same species.

Population cycle a regular variation of a population's size with time, often, although not always, associated with the availability of food.

Predator an animal which feeds upon populations of other animals (the prey); sometimes also used of herbivores, where the plant is the prey.

Productivity the amount of weight (or energy) gained by an individual, a species or an ECOSYSTEM per unit area per unit time.

Reproduction the process whereby populations increase in size.

Resources the basic requirements of an organism, including water, food energy and minerals.

Scavenger an animal which relies upon other animals to kill or collect food and then takes advantage of the unwanted remains.

Seasons the changes in climatic conditions associated with the changing angle of the sun in relation to the Earth's surface at various latitudes.

Sensory organs those organs of a plant or animal which are stimulated by the physical factors of the environment, such as light and heat.

Specialization the evolutionary development of a species leading to narrow limits of tolerance and a restricted role (niche) in the community.

Speciation the process by which new species evolve as a result of NATURAL SELECTION operating on genetically isolated populations.

Species a group of organisms which is distinct in being genetically different and usually isolated in its breeding from other organisms.

Stability in an ecosystem, the achievement of equilibrium and the development of resistance to change.

Stimulus an environmental factor which produces a response in an organism.

Succession the development and maturation of an ecosystem which eventually stabilizes in the climax.

Symbiosis an interaction between two species which is mutually beneficial.

Terrestrialization the process by which an aquatic habitat gradually becomes drier as infilling and peat development take place, eventually becoming a terrestrial (ie land) habitat.

Territory an area of land occupied by a single animal or a group of animals and actively defended by them, often by acts of display.

Tolerance limit the extreme value of a physical factor beyond which the species cannot survive.

Trophic level the feeding status of an organism, such as plant, HERBIVORE, CARNIVORE, and top carnivore.

Vegetation formation the major vegetation types of the world, defined by LIFE-FORMS, eg coniferous forest, temperate grassland, tundra, desert.

Water availability the ease with which water can be obtained from the soil by a plant.

Wind currents the pattern of air mass movements resulting from the uneven heating of the Earth's surface, coupled with the Earth's spin.

Zoogeographic region the division of the world into geographical areas on the basis of the taxonomic groups of animals present.

INDEX

A **bold number** indicates a major section of the main text, following a heading. A single number in (parentheses) indicates that the animal name or subjects are to be found in a boxed feature and a double number in (parentheses) indicates that the animal name or subject are to be found in a spread special feature. *Italic* numbers refer to illustrations.

A

B

C

D

U

V

W

X

Y

Z